There is often confusion over the meaning and usage of terms such as efficiency, economy, effectiveness, optimization and perfection in biology. This book defines and discusses these concepts within a broad evolutionary perspective and considers how evolutionary pressures can affect the economy and efficiency of animals. Chapters consider biomaterials, skeletal systems, muscular function, aquatic and terrestrial locomotion, respiratory and cardiovascular systems. The result is a book of interest to all biologists, and particularly to those working in the fields of comparative physiology and evolutionary biology.

Efficiency and economy in animal physiology

Efficiency and economy in animal physiology

Edited by

Robert W. Blake

Department of Zoology,
University of British Columbia,
Vancouver

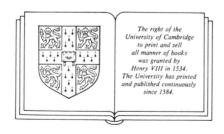

The right of the
University of Cambridge
to print and sell
all manner of books
was granted by
Henry VIII in 1534.
The University has printed
and published continuously
since 1584.

CAMBRIDGE UNIVERSITY PRESS

Cambridge

New York   *Port Chester*   *Melbourne*   *Sydney*

Published by the Press Syndicate of the University of Cambridge
The Pitt Building, Trumpington Street, Cambridge CB2 1RP
40 West 20th Street, New York, NY 10011-4211, USA
10 Stamford Road, Oakleigh, Victoria 3166, Australia

First published 1991

Printed in Great Britain at the University Press, Cambridge

British Library cataloguing in publication data

Efficiency and economy in animal physiology.
1. Animals. Physiology
I. Blake, R. W.
591.1

Library of Congress cataloguing in publication data

Efficiency and economy in animal physiology / edited by Robert W. Blake.
p.   cm.
Based on the proceedings of a symposium sponsored by the American
Society of Zoologists held in San Francisco, December 27th–30th, 1988
Includes index.
ISBN 0-521-40066-X
1. Physiology, Comparative–Congresses. 2. Evolution (Biology)–Congresses.
3. Adaptation (Biology)–Congresses. I. Blake, R. W. (Robert William), 1954-
II. American Society of Zoologists.
QP33.E44   1991
591.1–dc20 91-9215 CIP

ISBN 0 521 40066 X hardback

CONTENTS

CONTRIBUTORS

J. E. A. BERTRAM *Department of Organismal Biology and Anatomy, University of Chicago, Chicago, Illinois 60637, U.S.A.*

A. A. BIEWENER *Department of Organismal Biology and Anatomy, University of Chicago, Chicago, Illinois 60637, U.S.A.*

R. W. BLAKE *Department of Zoology, University of British Columbia, Vancouver, V6T 2A9, Canada*

T.L. DANIEL *Department of Biology, University of Washington, Seattle 98195, Washington, U.S.A.*

R. J. FULL *Department of Zoology, University of California at Berkeley, Berkeley, California 94720, U.S.A.*

C. GANS *Division of Biological Sciences, University of Michigan, Michigan, 48109, U.S.A.*

J. M. GOSLINE *Department of Zoology, University of British Columbia, Vancouver, V6T 2A9, Canada*

D. R. JONES *Department of Zoology, University of British Columbia, Vancouver, V6T 2A9, Canada*

G. V. LAUDER *Department of Ecology and Evolutionary Biology, University of California at Irvine, Irvine, California, 92717, U.S.A.*

W. K. MILSOM *Department of Zoology, University of British Columbia, Vancouver, V6T 2A9, Canada*

C. J. PENNYCUICK *Department of Biology, University of Miami, 33124, Coral Gables, Florida, U.S.A.*

PREFACE

This book is based on the proceedings of a symposium sponsored by the American Society of Zoologists, held during the society's San Francisco meeting (27–30 December 1988). Arguably, most physiological symposia focus, with varing degree of specificity, on issues within a classical division of the subject. This traditional approach has been, and continues to be, worthwhile and fruitful. However, the impetus here was to take a general theme (efficiency) and explore it broadly from a variety of perspectives and fields. The general consensus of opinion of the participants and audience at the San Francisco meeting was that the symposium was successful and worthy of publication. The resulting book consists of ten chapters (contributed by the symposium speakers) which vary somewhat in depth of coverage, length, style, and organization. This reflects a conscious decision on my part not to strive for uniformity through 'heavy handed' editing. A brief outline of the book's structure and content follows.

Chapter 1 (C. Gans) establishes definitions for a variety of terms (e.g. adequacy, efficiency, optima, perfection) that are employed by some authors in subsequent chapters. Gans discusses these concepts in the context of evolutionary biology, bearing on adaptation, development, and speciation. This issue is returned to in Chapter 10 (G.V. Lauder) where it is argued that concepts of efficiency may be used in integrating the discipline of physiology with historical biology. In Chapter 2 (R.W. Blake) the influence of the choice of formalism, interpreting high and low values, and the relevance of laboratory results to field situations are considered for efficiency criteria in physiological systems. Blake points out that some constraints on efficiency are set by physical and biological factors,

and this is explored more fully for the case of aquatic propulsion by
T.L. Daniel in Chapter 6. Daniel discusses thrust generation in swimming organisms of different size, morphological design, and kinematics
in relation to efficiency for steady and unsteady motions. Physiological limits to performance associated with muscle function, and energy
storage and dissipation in muscle are also discussed. C.J. Pennycuick
(Chapter 3) relates muscle efficiency to contraction frequency, and in a
novel analysis derives muscle efficiency from Huxley's equations for the
sliding filament model. In Chapter 4 J.M. Gosline reviews efficiency criteria for biomaterials. He focuses on elastic efficiency (resilience) showing that this parameter need not be maximized for effective function in
some cases. The efficiency of skeletal design is discussed by A. Biewener
and J. Bertram in Chapter 5. They point out that efficiency (defined as
strength per unit mass) is only one of a variety of factors necessary to
evaluate the performance of skeletons. The importance of historical constraints on skeletal form and function are also discussed. Complementing
Daniel's review of efficiency criteria for aquatic locomotion (Chapter 6),
R. Full considers the efficiency of terrestrial movement in Chapter 7. He
reviews a variety of mechanical efficiency criteria and emphasizes the
importance of considering other factors (e.g. endurance, stability, acceleration) in evaluating the performance of terrestrial locomotion. In
addition, the economy (cost of transport) of movement is discussed in
relation to body form. R. Full shows that whole animal mechanical
efficiency is highly variable and not simply equal to the efficiency of
isolated muscle. Chapters 8 and 9 deal with efficiency criteria for the
respiratory (Milsom, Chapter 8) and cardiovascular (Jones, Chapter 9)
systems. Milsom shows that the overall efficiency of the respiratory systems is a complex function of the anatomy of the gas exchange organs,
mechanics of the respiratory pumps and their pattern of ventilation. In
the final chapter, G. Lauder (Chapter 10) emphasizes the utility of efficiency criteria in understanding the evolution of physiological systems
when measured relative to an outgroup clade.

On behalf of all of the contributors to this book I would like to thank
the American Society of Zoologists and the National Science Foundation
for their financial support. Thanks are due to Mr A. MacDonnell and
Mr H. de la Cueva for their assistance in editing the manuscript. I would
also like to thank Dr Robin Smith and all at Cambridge University Press
involved in the production of this book.

Robert W. Blake

Vancouver

1

Efficiency, effectiveness, perfection, optimization: their use in understanding vertebrate evolution

C. GANS

1.1 HISTORY

Since before Aristotle, naturalists have noted that organisms are more or less matched to the environments they occupy. They also noted imperfections in that some organisms, individuals and species appeared obviously mismatched to the environment they then occupied.

Around the end of the eighteenth and start of the nineteenth centuries, these observations led to the wide array of hypotheses in Idealistic Morphology. Underlying many of these views was the idea that tissues and organisms were trying to express an innate pattern, in some schools referred to as the archetype. It was assumed by some students that the perceived world actually represented but a variable expression of this innate plan. Some species represented a closer fit to the underlying pattern and with this gave a better indication of the nature of this archetype. Hence, a major task of comparative biology was the deduction of the true archetype from the diversity of surviving species.

The theory of natural selection and its corollaries has provided a more appropriate explanation both for the kinds of environmental matching ... and for the seeming degrees of mismatching to the environment. Still, for various reasons, some biologists and philosophers even today keep searching for alternate explanations for this matching and mismatching. Alternative hypotheses (with which I do not agree) for the existence of mismatching are, for instance, the Spandrels of Saint Mark hypothesis, according to which adaptation (curiously defined) does not derive from natural selection. Another is the concept of evolution by punctuated events in which an array of individuals or species exemplifying different phenotypic states arises and only those lines that encounter an environmental site matching their particular condition will survive. Alternative hypotheses for the existence of matching included the concept (often ascribed to Cuvier) that one could reconstruct an entire fossil skeleton by informed analysis of a single bone. A more recent version is seen in the claim that study of its morphology might be sufficient to reconstruct the habits of a poorly known or fossil species. The underlying assumption is that matching should allow derivation of function from structure.

Both the terms matching and mismatching imply extremes of a standard for comparison and its characterization has been the task of this symposium. My viewpoint on this issue overlaps some of the following accounts and is based on (or biased by!) much watching of animals in laboratory and field. It is offered here as a set of standards for comparison applicable to general accounts of this topic, rather than as a set of new discoveries. The mention of vertebrate evolution is incidental to the general argument. Hopefully it will be found correct or at least stimulating for the reevaluation of such topics.

1.2   QUESTIONS AND DEFINITIONS

Terms, such as efficiency, compare at least two states; correctly used they should characterize a continuum. The need for such comparison leads to the fundamental consideration of what we should expect to see. Adherents of the theory of evolution by natural selection must then consider the question of what it is that natural selection may be likely to generate. The answer should lead to a state in which one can establish plausible predictions that may then be subjected to falsification.

However, this approach requires some standardization of the terminology used. To the extent that old terms are being used in a new framework, their existing definitions need to be considered. Such definitions are taken up here so that they may be reconsidered in terms of a biological framework.

1.2.1   Efficiency

Efficiency is a measure of performance relative to a physical or biophysical process or law. It always characterizes an actual performance, relative to an ideal or perfect level. It is independent of purpose and must be applied to a single process at a time. Examples might be the amount of thermodynamically determined heat now existing in body A that can be transferred to body B. Alternatively, the efficiency might characterize the fraction of the amount of heat in body A that can be transformed into mechanical work. In a different sense, it may refer to the number of molecules required to activate an olfactory organ.

1.2.2 Effectiveness

Effectiveness represents a performance directed toward a particular purpose and registers the degree to which this is achieved. It does not necessarily involve a chemical or physical limit (as does efficiency), but rather may involve a combination of physical and other traits. An example for its use would be in evaluating the design of lever system or a gear train intended to generate a particular force or velocity. Another example would be the evaluation of a linkage pattern incorporated in the limb or an animal, facilitating its ability to gallop or to manipulate its food.

1.2.3 Perfection

Perfection represents the best state that is conceivable. It would thus be equivalent to an efficiency of 100%. Obviously, perfection is only pertinent to a particular physical process, i.e. to heat transfer or force transmission. It is obviously the abstract Platonic ideal and, as will be seen below, is hardly pertinent in a biological situation unless one can exhaustively evaluate the mechanically or chemically dissected organism.

1.2.4 Optimum

The optimum represents the best state, but as qualified by a set of limiting (constraining) circumstances. Any claim for an optimum must specify what is being optimized as well as the limiting circumstances which are presumably operative. In the absence of such specification, a claim for optimization only expresses congruence of the observed with the presumed environmental demand without quantifying the closeness of the match or magnitude of selection.

Optima are often applied to circumstances in which there are several partially or completely conflicting factors so that modification of one affects the others. The optimum then represents a condition which is defined as the best overall for the prescribed task.

1.2.5   Adequacy (sufficiency)

Adequacy is a state that meets minimum conditions. These conditions obviously need to be specified in any actual application and the term hence is relative to these. Adequacy does not demand any level of efficiency or effectiveness; indeed it is independent of these. The relation of effectiveness and adequacy may be exemplified on the basis of the record of a college team. The effectiveness of the team may be seen in terms of the number of goals scored or blocked, runs earned or similar statistics. However, for any contest, there will be a victory, representing a terminal datum, based on the team that achieved the highest score. For purposes of the seasonal championship, adequacy only implies being better than was the opposition at that moment. In short, natural selection acts to produce adequate, not perfect, results.

1.2.6   Improvement

Improvement refers to two or more events separated in time. Improvement implies that a particular characteristic or process became modified during this interval and the modification increased its relative efficiency or effectiveness. Improvement also implies that the initial state was imperfectly matched to a purpose and that the modification increased or exceeded the initial level of performance. Improvement often reflects intrapopulational selection or response to changing competitors, prey or predators.

1.2.7   Patterns in evolutionary biology

It is next useful to ask again how these terms apply to the description of biological phenomena (Fig. 1.1). Do the processes of adaptation and speciation allow predictions about the level of effectiveness likely to be observed? Such considerations may be analyzed by a simple reconsideration of processes of evolutionary change.

We know that any presently surviving species must be sufficient for the environment or environments it now occupies. This very obvious statement has sometimes been claimed to result in an adaptationist tautology; however, it remains true. However, persistence of survival can be assessed independently by physiological or structural criteria, for instance, cold-hardiness, tenacity of adhesion vs. maximal wave force. If we wish to make this statement for a form occupying an environment

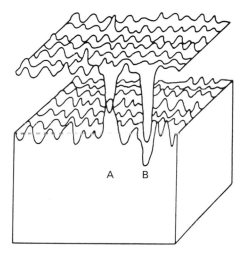

Fig. 1.1. Sketch to permit visualization of the relations between the multiple environmental demands imposed upon animals and the phenotypic state of individuals. The surface of the (lower) block is intended to indicate the multiplicity of components of the myriad related and independent aspects of the environment. The vertical dimension of each segment of the block represents the magnitude of that aspect. The (straight) horizontal line is adjusted to show the mean magnitude of that aspect and the complex surface shows the particular conditions encountered by a particular individual (at one moment in time). The complex top surface illustrates aspects of that individual's phenotype expressed in terms of its capacity to match the environmental demand during a particular interval of time (which may be its life span). [Note that it is impractical to map mean phenotypic response as its position will differ with the magnitude of what is sometimes called the 'factor of safety' for each aspect.] At point A, the momentary environmental demand is greater than the phenotypic capacity. Hence the element is likely to fail. The importance of the failure will depend on the significance of this and related phenotypic aspects to the life of the organism. At point B, the phenotypic aspect is insufficient to match the mean environmental demand. However, the actual demand that the individual encounters is less than this; hence the individual is adequate and the element does not fail. The general scheme indicates how most aspects of individuals are excessively constructed for much of their life history and that this 'excessive capacity' permits behavioral adjustments without failure.

that remains stable over one or more generations, the statement implies that individuals of all sexes and of all ontogenetic stages must be sufficient.

Obviously, the population of any species will show phenotypic variations. The concept of sufficiency does not imply that all phenotypic variants of a population will indeed be sufficient; it only implies that enough individuals will display sufficiency so that the species will be maintained. Whenever the species occupies variable or fluctuating environments, there will be complications, such as populational polymorphisms, but the same general considerations apply.

Every individual organism occupies a multidimensional volume within a parameter space representing aspects that have to be tolerated and those that may be exploited. Different parameter spaces will be occupied by individual members of a species at various times during their ontogeny. The axes of such a space are likely to specify an enormous number of phenotypic states. Examples might be: The maximum torsional strain in a midlevel humerus. The maximum deformation of a contour feather during loading (in the physiological range). The minimum turning radius while flying at a particular velocity. The minimum number of photons needed to activate a visual cell.

The limits of this parameter space are established by the conditions of the environment, both biological and physical. In the general case, each parameter changes continuously in time. Hence, individuals of a species that are marginal in one dimension may survive or not, depending on the time at which they live. This is another way of stating that sufficiency will be relative to the present environmental conditions testing the system.

The organism involved in such a system must grow, during which time their phenotype is transformed into a juvenile phenotype and this into that of the adult. As the transcription of any genotypic instructions is modified by the environment, which is in turn variable, the developmental process inevitably involves (environmentally induced) variations. These likely incur selection, the magnitude of which depends on the closeness of their fit to the particular limits of the volume of selective pressures. The closer the mean phenotype matches a particular limit, the greater the chance that some individuals of the population will be inadequate for that limit. However, an increased factor of safety will likely incur a greater cost in generation and maintenance of a phenotype matching or surpassing the particular limit.

More important is that there are other and often unpredictable factors limiting the matching of the phenotypic range to the environmental limits; all of these constrain the distribution of phenotypes. As the other factors may interact, they generate synergistic or interference effects, linear or otherwise. Examples of such sources of unpredictable noise in the process of phenotype specification are the pleiotropy of genes, the fact that characters generally have multiple roles, simultaneously or sequentially, and that genetic instructions must code for juveniles and adults, with the transitions involving further compromises. Such factors vitiate simple attempts to resolve the structure-function match, and to derive function from the structure of an individual.

1.2.8 The structure/role match

Hence, the phenotype of an individual does not permit simple extrapolation to its roles, those of its functions that are of significance to the organism. Not even the range or mean phenotypes of a species will allow accurate determination of most roles. However, we may assume that the importance of the role of a phenotypic aspect to the organism, and the cost of the aspect, will influence the tightness of the match to the limits of the parameter space.

All of this has been stated for a steady state. However, environments are never steady and individuals always test their limits, or conversely, environments always test the limits of organisms. Sometimes this testing lets the individual reach adjacent areas of the heritable character space. If their phenotype is incidentally at least sufficient for local maintenance, they may survive there as individuals. If enough such individuals do reach these zones, and if they are sufficient to achieve reproduction, this may lead to a shift of all or part of the original population into a modified character space. As the invaders need not be genetically typical of the original population, there is then the potential for a genetically different population, which in turn may change its phenotypic range further under the selective effect of the modified space. In essence, they will incur improvement toward a new set of limits potentially generating further transitional and intermediate states.

The phenotype, particularly of more complex organisms, also incorporates the capacity for degrees of behavioral adjustment in the use of its other aspects to match the conditions of a modified environment. This may complicate identification by the investigator of the roles shaping the organism. However, it may also permit closer matching of phenotype to particular limits.

1.2.9 Evaluation of performance

If these observations are correct, what level of performance should we expect to see in organisms? A significant fraction of the offspring of any mating should be adequate for the limits of their environment.

This fraction will reflect the reproductive rate, i.e., it will be smaller in a halibut or a sea turtle (with offspring counted in the thousands or hundreds) than in giraffe or humans (which produce one or two at a time). In each such case one may be able to identify one or more limiting conditions (i.e. axes of the parameter space) in which a subgroup of the population will be adequate under some but inadequate under other environmental circumstances.

The adequate fraction of offspring likely may be greater for physiological aspects than for the avoidance of predators. In short, many individuals taken by predators may be expected to have been otherwise adequate. However, in other species the presence of nutrients, or water, or heat, or settling sites might be as critical for (or associated with) the ability to avoid predation.

It is likely that the categories glibly defined are interactive. For instance, predation is likely to be successful against a subset of the physiologically less adequate or otherwise marginal, even though the 'inadequacy' is independent of the predator's success. This suggests that definition of the limiting condition must be handled carefully. For instance, simply affecting survivorship of a population by eliminating predation at one stage may have unexpected consequences. Identification of the limiting circumstances is sometimes possible for a particular condition (i.e. axis of the parameter space). Among other cases, this includes the situations in which 'adaptation' for a particular (presumably limiting) role seems obvious. One would expect these cases to have the following characteristics:

1.   The role is very important to the organism and may differ from performance factors of related forms.
2.   The role has persisted for a substantial time in the history of the clade (so that it is generally expressed and possibly associated negative factors are being buffered).

Tests will be needed to monitor the resulting congruence or the closeness of fit between environment and role. Sample tests might be the following:

1.   What is the range of variability for the characteristic. How close is the mean, 80% of population or 95% of population to the putative environmental limit? What is the distribution - normal, skewed, etc.?
2.   Are these values and distributions different for other aspects?
3.   To what extent are these values heritable?

4.   What kinds of environmental factors affect population phenotype and what is the direction of such effect? How variable are these factors, i.e. what is time constant of change relative to the life span of the organism?

1.2.10   Terminology revisited

We may now return to the question of how these terms fit into evolutionary biology. The term efficiency had best be set aside or confined to physiology. It is rarely usable, only pertinent to physical laws and by implication to a single aspect at a time, an aspect which is compared to perfect state.

The term effectiveness should be used for most biological systems to which efficiency is often misapplied. It represents a general measure of performance in a well-defined role. Optimality or optimization represents the best state (or process) from among a set of limiting circumstances. Optimality does not necessarily quantify closeness to an ideal; however, this may be specified. As the particular environmental framework establishes the relative needs, the available parameter space must be specified in any definition. Hence, optimality remains a relative term, often pertinent only to express what might be called the degree of adequacy. We have seen that optimum does not imply any independent level of efficiency or effectiveness. Even less should there be the implication that levels of optimality will be equivalent for the steps or stages of a physiological or behavioral process. However, it may be useful to introduce the term Congruence here, defined as the degree (closeness) of matching of phenotypic and environmental parameter spaces. A degree of congruence may reflect that selection upon heritable characters is indeed occurring.

Hence, congruence may document current adaptedness, particularly if experimental modification of the factor alters fitness. The concept of congruence makes it possible to measure improvement over time; however, the nature of the task or role must first be defined.

1.2.11   Vertebrate evolution

Courses in the Comparative Anatomy of Vertebrates often derive their justification from the possibility of a sequential arrangement of the vertebrate 'classes'. This is reflected in the occurrence of paraphyletic groups that raises such ire among some taxonomists.

One byproduct is the tendency to claim increases in efficiency or performance with changes from fishes to reptiles to mammals. However, only some characteristics, such as gas exchange, water balance and perhaps locomotor patterns do show general trends; even in these, there are exceptions and repeated attempts at special solutions, for instance, for return to the oceans or conquest of the air. Other characteristics lack continuing trends; for instance, the vertebrate feeding mechanisms are spectacularly opportunistic, as are internal fertilization and vivipary. If we look at energy metabolism, we see the reverse to a long-term trend in effectiveness. The shift from cephalochordates to vertebrates was marked by an enormous increase in metabolic scope without obvious increase in the capacity to generate more energy or the efficiency of that generation. Rather, the increased scope allowed the resulting vertebrates to expend five or more times as much energy because possession of the scope allowed them to increase the quantity and quality of ingested food. Similarly, the transition from ectothermy to endothermy generated a ten times more profligate life style that could be supported because previously unavailable nutrients could be harvested. Efficiency or effectiveness was not the determining factor; the animals used what was available to be harvested. Thus, horses and elephants high grade their food, extracting a limited fraction from a very large potential harvest, rather than extracting a greater fraction from a potentially smaller resource. Hence, increase in absolute values can be just as significant as increase in relative effectiveness.

1.3   CONCLUSIONS

The key question of this symposium has been at what level different aspects of the phenotype match the limits of the environment, and what are the concepts that such comparisons imply. Resolution requires understanding of the current role or roles. Also of the time base on which comparisons are made. The main problem to achieving resolution remains that environmental conditions vary unpredictably and that organisms are intrinsically variable and respond opportunistically. They use their existing phenotypes in varied, often novel ways. Species can tolerate varying phenotypic states because phenotypes only need to be adequate. Analysis of groups of organisms in a systematically comparative framework can often establish direction of past modifications. Most of all, we need comparisons of the extent of congruence of the phenotype-role match. These will provide estimates of the degree of optimization of a particular system. Such comparisons must be directed simultaneously at multiple characteristics at multiple species in a phyletic framework.

1.4   ACKNOWLEDGEMENTS

I thank the organizer for the invitation to participate in this symposium and Robert Dudley for comments. A grant from the Leo Leeser Foundation provided support for preparation of this manuscript.

2

On the efficiency of energy transformations in cells and animals

R.W. BLAKE

2.1 INTRODUCTION

Biologists study energy transformations in cells and organisms at various hierarchical levels, from biochemical pathway processing through to ecological energetics. The approaches employed to understand them are diverse and field dependent. The methods and approaches commonly used to study energy transformations in mitochondria are not those employed in animal locomotion. Assessments of the effectiveness and performance of energy transformations are often based on efficiency criteria. Arguably, many of the issues concerning the formulation, application and interpretation of efficiency values are of a general nature.

This paper explores six arbitrarily defined themes concerning efficiency. These themes (influence of the choice of formalism on numerical values, dangers associated with commonly expected results, interpreting high and low values, reconciling physiological and mechanical findings, the relevance of laboratory results to field situations, and scaling) are discussed with examples. Some remarks regarding evolutionary perspectives are made in a concluding section.

2.1.1 Numerical results may reflect the choice of formalism and its associated assumptions

Medawar and Medawar (1983, pp. 66–67) argue that the importance of definitions in biology is highly exaggerated, and that biology can proceed without the regard for clear definitions that is essential in mathematics. It may be that tight definitions are not required and/or possible in some areas of biology. However, the definitions and contexts of biological efficiency criteria must be clear. Numerical results reflect the nature of underlying theory. Examples from biochemistry and biomechanics illustrate this below. The context of the application is also important. A given efficiency parameter may refer to a stage of a process, an overall process, or a series of processes. An example from muscle physiology is used to illustrate this point.

There are a number of methods for assessing the efficiency of biochemical pathway processing. Criteria may be based on either equilibrium or non-equilibrium thermodynamic approaches. A number of common procedures assume that the reactions being considered are at or near equilibrium conditions. For example, the efficiency of coupling of oxidative phosphorylation is often calculated as the ratio of the standard Gibbs free energy of reactions. Rather than dividing the standard Gibbs energies of the reactions under consideration some authors focus on ATP yields. Efficiency criteria may also be based on the molar thermodynamic yield (net energy available in products from a mole of reactants) where efficiency is indicated by the percent completion of the reaction being considered. Cornish-Bowden (1983) gives examples of efficiency parameters based on equilibrium assumptions and notes that values of 40–60% are typical of many energy yielding pathways.

Recently there has been much criticism of efficiencies based on standard Gibbs free energy ratios and ATP yields because many important reactions may be displaced from equilibrium (Atkinson, 1977; Cornish-Bowden, 1983; Watt, 1986). Current opinion favours concepts of efficiency based on non-equilibrium thermodynamics.

Classical (equilibrium) approaches consider only net forces (chemical affinities), non-equilibrium criteria views the energy production of a series of coupled reactions in terms of net forces and flows (reaction rates). Stucki (1980) formulates a linear model of oxidative phosphorylation. The flows considered are the net flow of ATP and oxygen consumption and the forces are the phosphorylation potential and redox potential between electron accepting and donating couples. The degree of coupling of the reactions involved is described by a series of phenomenological coefficients associated with the flow terms. The efficiency of oxidative phosphorylation is given by the ratio of output power (phosphate potential multiplied by the net rate of ATP production) and input power (product of the net rate of oxygen consumption and the redox potential). The optimal efficiency for NADH dependent substrates is about 60% (Stucki, 1980; see Fig. 2.1).

Watt (1986) defines efficiency criteria valid for situations that are far from equilibrium states. Time dependent transient efficiency expressions are integrated over the course of reactions and combined with steady state expressions to give the overall metabolic processing efficiencies of both single reactions and reaction sequences. Figures for the efficiency of oxidative phosphorylation employed by physiologists (50–60%) are commonly based on inappropriate equilibrium approaches. Fortunately, in this case more legitimate methods give similar numerical results.

Locomotor performance is often assessed through a propulsive efficiency parameter, defined as the ratio of the power usefully employed in propelling an animal to the total power delivered to the propulsive structures. Interestingly, the concept of propulsive efficiency has not been widely applied in studies of terrestrial and aerial locomotion, but is central to most discussions of swimming. The example discussed below concerns undulatory swimming.

Undulatory swimming is the most common means of aquatic locomotion. The reasons for this evolutionary success are both hydromechanical and structural. Undulatory propulsion is relatively insensitive to changes in scale (represented by the Reynolds number R_e , a ratio of the inertial to viscous forces acting in the fluid). It occurs in bacteria and blue whales, which operate at Reynolds numbers of about 10^{-6} and 10^7 respectively. There is no evidence to suggest that there are any structural limitations, both segmented and non-segmented animals are propelled by undulatory movements.

The propulsive efficiency of undulatory locomotion can be determined from hydromechanical theory. Resistive theory (e.g. Taylor, 1952; Gray and Hancock, 1955) is applicable at low Reynolds numbers (< 1) where viscous forces dominate and the linear resistance law applies. At higher Reynolds numbers the quadratic resistance law applies and reactive (inertial) theory is appropriate (Lighthill, 1975; Blake, 1983). Reasonable estimates of swimming thrust, power and efficiency require a model relevant to the Reynolds number of the system. In addition, the assumptions of the model in question must be met. Webb (1975) has shown that resistive hydromechanical models applied to systems at higher Reynolds numbers where inertial effects are dominant will tend to underestimate propulsive efficiency. Difficulties should not arise for high and low Reynolds number (> 1000 and < 1 respectively) situations where the choice of model is clear. Unfortunately, there are no obvious approaches to systems that operate at intermediate Reynolds numbers (1–1000). This problem is important because a great many animals swim at intermediate Reynolds numbers. The efficiency of undulatory propulsion is discussed further by T. Daniel in Chapter 6.

Below, an example from muscle physiology serves to emphasize the importance of establishing clear definitions and contexts for efficiency values. Care must be taken in interpreting the context of values as they may be concerned with one or more of the following; the efficiency of the contractile process per se, of the contractile process and associated ion pumping, the overall efficiency of either maintained isotonic contractions, or complete cycles of contraction and relaxation.

Kushmerick and Davies (1969) report values for the efficiency of the contractile process plus ion pumping based on determinations of overall mechanical efficiency, partial molar enthalpies ΔH and free energies ΔG of the relevant reactions. Efficiency is expressed as the product of the overall mechanical muscle efficiency and $\Delta H/\Delta G$. Values for frog sartorii at $0^o C$ shortening at constant velocity of about 66% were found. Subtracting out the ATP associated with ion pumping, values close to 100% are implied for the efficiency of the contractile process per se. Biophysicists measure the power output and mechanical efficiency of excised striated muscle by determining the work done ΔW and the heat produced ΔH during maintained isotonic contractions. The experiments may involve shortening at a constant speed (e.g. Kushmerick and Davies, 1969) or load (e.g. Hill, 1964). Mechanical efficiency is expressed as $\Delta W/(\Delta W + \Delta H)$. Maximum values of about 40–45% are found when the muscle contracts with about 50% of its maximum force and about 20% of its maximum shortening speed. Unfortunately, some authors have employed these figures in discussions of locomotion involving cycles of contraction and relaxation. For maintained isometric contractions ΔH has two components, activation and shortening heat. A relaxing muscle generates a heat of relaxation that is about equal to the sum of activation and shortening heat. Consequently, the mechanical efficiency of cycles of contraction and relaxation is about one half that of maintained contractions.

In situ values of muscle efficiency are commonly based on whole body ergometry and calorimetry measurements. Steady state aerobic exercise on bicycle ergometers give values of about 20–30% (e.g. Pugh, 1974; Gaesser and Brooks, 1975). There are a number of factors that influence the results, among them, the type of bicycle ergometer used, work load, rate and duration of the exercise, training and motivation of the subjects etc. However, results may also depend on exactly how muscular efficiency is defined. There are four definitions of muscular efficiency in the exercise physiology literature: gross efficiency (work done/total energy expenditure), net efficiency (work done/work done above that at rest), work efficiency (work done/work expended beyond that done in cycling without a load), and delta efficiency (increment in work performed above a previous work rate/increment in overall work above that at a previous rate). Gaesser and Brooks (1975) found either a linear or slightly exponential relationship between caloric output and work rate implying either a constant or decreasing efficiency with increasing work rate. Only the delta efficiency gave this result and Gaesser and Brooks (1975) suggest that it be employed in preference to other measures. More detailed accounts of muscle efficiency are given by C. Pennycuick and R. Full (Chapters 3 and 7 respectively).

2.1.2 *Beware the common expectation*

Values of efficiency are often assumed from previous work. For example, workers in animal locomotion commonly assume values for muscle efficiency based on results from muscle and exercise physiology. Below, examples from terrestrial and aerial locomotion indicate that even a large data base derived from a variety of approaches may not provide the basis of a generally applicable rule.

Since the work of A.V. Hill and others on isolated muscle preparations it has become axiomatic to assume that the mechanical efficiency of repeated cycles of contraction of working muscles is of the order of 20–30% (see Pennycuick, Chapter 3). Results from bicycle ergometry (e.g. Pugh, 1974; Gaesser and Brooks, 1975) support the work on excised muscles. However, recent work on running in terrestrial vertebrates and hovering flight in certain insects casts doubt on the general applicability of the 20–30% rule of thumb for in situ muscle efficiency.

Cavagna *et al.* (1977) employ force platform, kinematic, and oxygen consumption data in studying the energetics of terrestrial locomotion in bipeds, quadrupeds, and hoppers. Muscle efficiency is expressed as the ratio of the power required to increase the mechanical energy of the centre of mass to the chemical power input to the muscles. On the basis of their results Cavagna *et al.* argue that in situ muscle efficiencies exceed the common expectation because of elastic strain energy storage in the system. In addition to considering the kinetic energy of the body relative to its centre of mass (external work), C.R. Taylor and his associates consider the kinetic energy associated with the limbs (internal work), defining efficiency as the total mechanical power output divided by the

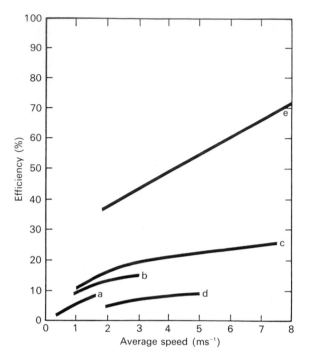

Fig. 2.1. Muscular efficiency (total mechanical work divided by metabolic energy input) is plotted against running velocity for: a, 43 g painted quail; b, 107 g chipmunk; c, 5.0 kg dog; d, 6.4 kg turkey; and e, 70 kg human. Based on Heglund *et al.*, 1982b.

total metabolic power input. Studies have focused on the importance of elastic energy storage (indicated by efficiency) in relation to body mass and speed (e.g. Taylor *et al.*, 1982; Fedak *et al.*, 1982; Heglund *et al.*, 1982a,b). Muscular efficiency increases with increasing body size and running speed (Fig. 2.1), reflecting the importance of elastic energy storage in larger, faster animals (see R. Full, Chapter 7 Fig. 7.1). Maximum estimates of efficiency are of the order of 70% for running humans. For elastic energy storage to be significant, isometric force must be generated by the skeletal muscles and it may be that a major portion of the energy cost of vertebrate terrestrial locomotion is associated with this.

Ellington (1984) has determined the power output in hovering flight for a variety of insects from a biomechanical model and combined this information with literature measurements of the total and basal metabolic

rates of insects to determine the mechanical efficiency of the flight muscles. Mechanical power output was divided by active metabolic power
to give muscle efficiency. Assuming perfect elastic energy storage values
ranged from 5–10%. Values close to the common expectation of 20–30%
are obtained if it is assumed that the flight muscles do all the work necessary to accelerate the wings. However, Ellington points out that the
absence of elastic energy storage would imply an unusually high power
loading (power output per unit weight) from the flight muscles. Any-
way, elastic energy storage mechanisms are present (Weis-Fogh, 1960),
and it would be unreasonable to suppose that they do not function. The
common expectation for the efficiency of striated muscle is not surprising
given its structural uniformity (e.g spacing, arrangement and dimensions
of myosin and actin filaments) and broadly similar mechanical properties (strain, maximum isometric stress, etc.). Given this structural and
mechanical similarity it may be tempting to suggest that the in situ locomotor performance of muscle is likely to depend on other things, such
as the amount of muscle involved in a particular task, its gross anatomical arrangement and fibre geometry (parallel, pinnate), and type (slow,
fast, etc.) However, if mechanical efficiency proves to be a variable characteristic of working muscles, then it may be important in determining
the relative energetic cost of locomotion in different animals.

2.1.3   High and low values

Deeming an efficiency value high or low is not always straightforward.
Examples from swimming and flight are employed to explore some of
the issues concerning the assessment of values. They serve to illustrate
that low values may not reflect poor performance per se. Rather, low
values may reflect inherent biological and/or physical constraints.

Sometimes arbitrary 'cut-off' criteria are applied to evaluate efficiency.
For example, Lighthill (1970) suggests a division of undulatory swimmers
that operate at higher Reynolds numbers between forms characterized
by good and bad hydromechanical shapes, corresponding to a propulsive
efficiency of about 50%. This distinction is worthwhile for assessing
the relative swimming performance of active, steady swimming, pelagic
fish. However, for broader comparisons, involving undulatory propulsion
in forms characterized by different size and swimming behaviours this
procedure is not useful.

Most invertebrate undulatory swimmers are less efficient than fish.
However, many invertebrates swim at lower Reynolds numbers and the

propulsive efficiency of undulatory propulsion increases with increasing R_e. At lower Reynolds numbers viscous forces are important and resistive hydromechanical models apply. In the resistive theory of undulatory propulsion maximum propulsive efficiency depends on the ratio of the tangential and normal force coefficients (K_T and K_N respectively) of the oscillating parts.

Maximum propulsive efficiency is given by $(1 - (K_T/K_N)^{1/2})^2$. Maximum propulsive efficiency for flagellates ($R_e \approx 10^{-3}$) is about 5%, for nematodes ($R_e \approx 1$) it is 0.15–0.25, and for smooth-bodied polychaetes ($1 < R_e < 100$) values of the order of 0.3–0.6 are found. This trend of increasing propulsive efficiency with increasing Reynolds number reflects the dependence of the ratio K_T/K_N on R_e. So, there is a physical constraint on the maximum possible propulsive efficiency of undulatory swimmingat low Reynolds numbers and it is not legitimate to suggest that undulatory swimmers operating at low Reynolds numbers perform badly relative to those at higher R_e. Of course, differences in propulsive efficiency for animals as diverse as flagellates, nematodes, polychaetes and fish will also depend on a variety of other things (morphological differences, exact kinematics, etc.). Nevertheless, the possibility of extrinsic physical constraints should be considered when evaluating and comparing efficiency values.

Weis-Fogh (1972) defines an aerodynamic efficiency parameter and employs it to compare the performance in hovering flight of a hummingbird *Amazilia* and fruit fly *Drosophila*. The mass specific minimum power (an estimate of the induced power corresponding to a steady downward momentum jet, expressed in Watts per Newton) is divided by the mean mass specific aerodynamic power actually expended to give efficiency. Values of 0.52 and 0.27 were estimated for *Amazilia* and *Drosophila* respectively. Weis-Fogh (1972) suggested that the difference between the values can be attributed to the relatively low Reynolds number at which *Drosophila* operates. Accepting this, we have another case of efficiency reflecting a physical constraint rather than any fundamental difference in performance per se. Many snakes swim in the whole body (anguilliform) undulatory mode and are characterized by propulsive efficiencies that are significantly less than fish swimming the same way. This may be attributed to the relatively 'poor' hydromechanical shape of snakes (Lighthill, 1969). Among other things, posterior lateral compression is required for good performance in the anguilliform mode and many snakes are of roughly circular cross-section. A relatively flattened ventral surface is required for effective terrestrial locomotion. Arguably, the form of many snakes reflects a design compromise for effective swimming and terrestrial movement. Interestingly, sea snakes are characterized by posterior lateral compression.

The propulsive efficiency of rough-bodied nereidiform polychaetes is significantly less than that of smooth-bodied forms. Taylor (1952) showed that for rough-bodied worms K_T may exceed K_N, and that under this condition forward propulsion is possible on the basis of a retrograde wave. For *Nereis* $K_T/K_N \approx 1.5$ giving a propulsive efficiency of only 5%. Given that rough-bodied polychaetes are far less efficient than smooth-bodied forms it is unlikely that parapodia evolved as a swimming adaptation. Clark (1964) suggests that parapodia allow for locomotion on soft substrates. In this light, the relatively low value for the propulsive efficiency of nereidiform polychaetes may be viewed as reflecting the conflicting demands of different locomotor modes.

Blake (1986) compares the efficiency of paddling propulsion in the Angelfish *Pterophyllum* and water boatman *Cenocorixa*. The propulsive efficiency of the Angelfish is only half that of the water boatman. The thrust generated by a paddle is proportional to water density, relative velocity squared, blade area, and a force coefficient. Relative velocity increases linearly from the base to the tip of the paddle. For good performance, most of the area of the paddle should be distally located to coincide with high values of relative velocity. The pectoral fins of the Angelfish are broad based. The in board area of the fins does not generate much thrust and may produce negative flow interactions (interference drag) with the body. The propulsive hind limbs of the water boatman consist of a central jointed spar fringed distally with setae, most of the limb area is outboard and contributes usefully to thrust development. It could be argued that the water boatman paddle is hydrodynamically superior to the Angelfish design and that this is reflected in the efficiency values. However, the Angelfish employs its pectoral fins in an undulatory mode when hovering. This requires that the fin rays be separate at their base. The Angelfish pectoral fin reflects a compromise in design for effective hovering (undulatory propulsion) and forward movement (paddling). The water boatman hind limb is not similarly constrained in design.

Physical constraints and functional design compromises associated with different locomotor requirements and/or activities (e.g. feeding) must be borne in mind when interpreting efficiency values. However, engaging in more than one type of locomotion does not necessarily imply decreased efficiency. Some forms can change shape when the locomotor mode is changed. For example, leeches show an approximately circular cross-section when crawling, but contract dorso-ventral muscles to produce a flattened form when swimming (Clark, 1964).

2.1.4   Efficiency and economy: a paradox?

The economy of movement is commonly expressed as the cost of transport which is determined by dividing the energy equivalent of the rate of oxygen consumption by the product of speed and body weight. Values are given for a characteristic speed (e.g. maximum sustainable aerobic speed), and may include or exclude the standard metabolic rate. Tucker (1970) has shown that the cost of transport is directly proportional to the resistance to movement and inversely proportional to efficiency and weight. For animals of a given weight, moving through the same medium, in the same way, at the same speed, differences in the cost of transport should reflect differences in efficiency. High efficiency should imply low cost of transport and vice-versa. There are examples of this. Man is an inefficient swimmer and the cost of transport for this activity is high (Schmidt-Nielsen, 1972). However, the relationship does not always hold.

Plots of the cost of transport against body weight show that a single regression line can accurately predict the cost of transport of forms at given body weight moving in a particular way (Schmidt-Nielsen, 1971; Tucker, 1975). Schmidt-Nielsen (1971, 1984) plots the cost of transport of salmonids in relation to size (weight). He notes that in addition to accurately predicting the cost of transport for salmonids, the resulting regression line also gives good estimates for other fish that differ markedly in morphology and swimming style from salmonids (Fig.2.2). It is not surprising that salmonids fall on the same line. Their body plan, mode of swimming (subcarangiform), muscle and support systems (respiratory, cardiovascular) are essentially the same and Webb *et al.* (1984) have shown that their propulsive efficiency at maximum sustainable cruising speeds is not size dependent. Conversely, it is not clear why forms as diverse as *Thymallus, Coregonus, Tilapia, Lepomis* and *Lagodon* fall on or near the salmonid curve. Given the extent of the morphological and kinematic differences between these fish and salmonids, significant differences in cost of transport and propulsive efficiency would be expected on biomechanical grounds (e.g. Webb, 1975; Blake, 1983). However, physiological (respirometric) determinations suggest that morphological

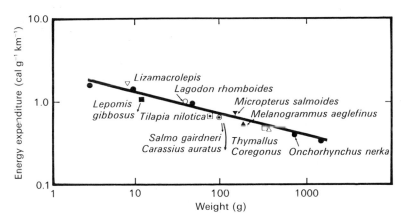

Fig. 2.2. Cost of transport in relation to body size for a variety of fish. Based on Beamish, 1978.

design and swimming mode are not important in determining the cost of transport.

Herreid (1981) compares data on the cost of transport of terrestrial arthropods with that on reptiles and mammals and concludes that all pedestrian animals have grossly similar cost of transport at given body weight regardless of their form of respiration, circulation, and number of legs. Peters (1983) compares the economy of movement in terrestrial birds, lizards, and mammals and notes that they can all be described by a single regression line. In Chapter 7 (Fig. 7.9), R. Full summarizes transport costs of terrestrial animals in relation to body mass. Again a single curve of negative slope adequately describes the metabolic cost of transport in relation to body mass (Chapter 7, Fig. 7.9 upper curve). The lower curve (Chapter 7, Fig 7.9) shows that the mechanical work associated with moving a unit mass a unit distance is essentially size independent. This implies a positive relationship between the efficiency of operation of the muscular and/or support systems with increasing size. The improvement of the efficiency of running with increasing size due to increases in elastic energy storage has already been noted (see Fig. 2.1). However, the lower curve (Chapter 7, Fig. 7.9) accounts only for the mechanical work associated with movements of the centre of mass (internal work). Among other things, the contribution of movements of the appendages etc. relative to the centre of mass (external work) in relation to mass must be assessed before a complete explanation for the scaling of cost of transport in terrestrial animals can be offered.

Reconciling physiologically obtained cost of transport data with biomechanical analysis is also problematic for flying animals. Based on conventional aerodynamic approaches, a plot of the total aerodynamic power against flight velocity generates a U-shaped curve. If basal metabolic costs and muscle efficiency are independent of speed, then a plot of total metabolic power input against velocity should also be U-shaped. Most studies however, have generated rather flat metabolic power input curves (e.g. see Berger, 1985, Fig. 3). This may reflect speed dependent variations in muscle efficiency and/or basal costs. Alternatively, the quasi-steady aerodynamic models used to determine power output may not be appropriate.

2.1.5   Scaling

Unfortunately, there is insufficient data in most areas to allow for scaling efficiency in terms of body mass and performance. Available information suggests that the scale dependency of efficiency on body mass ranges from weak to strong. For example, overall cardiac efficiency (rate of working of the heart divided by the work equivalent of its oxygen consumption) in mammals increases very slowly with increasing body mass (Schmidt-Nielsen, 1984). In contrast, in situ muscle efficiency of running in terrestrial vertebrates increases rapidly with increasing size (see Fig. 2.1). It is likely that in many cases the scaling of efficiency of a given process or mechanism will reflect a variety of interacting factors. For example, it is probable that the decrease in the cost of transport of animals with increasing body mass reflects improvements in the efficiency of locomotion and the support systems and does not reflect the influence of a single selective force.

2.1.6   Laboratory estimates may not apply in nature

Many studies of animal physiology leading to efficiency estimates are laboratory based and it is assumed that the values obtained apply in nature. Examples from locomotion, respiratory and cardiovascular physiology are employed to cast doubt on the validity of this assumption.

Laboratory experiments on the energetics of animal locomotion typically involve the investigator setting the mode and activity level of locomotion. Most work involves animals that are either forced or trained to keep station on treadmills, water or wind tunnels. Given levels of activity are maintained by individual animals for pre-determined periods.

Strictly speaking, measurements of power requirements, cost of transport and efficiency are only relevant to the particular movement pattern seen in the experiment. Recently a variety of energy efficient locomotor strategies have been described for fishes and birds. Mathematical models of given movement patterns for individuals and/or interactions of a group indicate substantial energy savings. Fish show a variety of well documented locomotor strategies, including intermittent swimming (burst-and-glide, burst-and-coast swimming in negatively buoyant and neutrally buoyant forms respectively), tidal stream transport, and schooling (see Blake, 1983, for a review). Birds engage in intermittent (bounding) flight (Rayner, 1977), soaring (Pennycuick, 1972) and formation flight (Lissaman and Schollenberger, 1970). It is likely that the cost of transport and efficiency of free-ranging animals is significantly different from laboratory values.

Many studies on respiratory physiology focus on the kinematics, mechanics, and energetics of breathing. The conditions for breathing at minimum work and the mechanical efficiency of respiration are commonly considered (e.g Otis *et al.*, 1950; Otis, 1954; Agostoni and Mead, 1964; Spells, 1969; Perry and Duncker, 1980; Milsom and Vitalis, 1984). Estimates of breathing efficiency are usually based on determinations of the mechanical work done and the total energy required by the respiratory muscles. Because of the technical difficulties involved in simultaneously measuring the mechanical work done and the oxygen demand of the respiratory muscles few studies have been done on subjects other than man. A key difficulty lies in assessing the oxygen cost of ventilation as a proportion of the overall energetic cost of activity. Given this, most information on the energetics and efficiency of breathing in vertebrates has been obtained from resting or anaesthetized animals. It is necessary to assume that the results gained from such preparations reflect the costs etc. incurred during movement. I know of no studies that directly assess the validity of this assumption. In addition, there is evidence to show that respiratory and locomotor movements are coupled.

Locomotor-respiratory coupling (correspondence of the frequency of respiratory and locomotor movements) has been documented in man (e.g. Asmussen, 1967), quadrupedal mammals (e.g. Bramble and Carrier, 1983), and bats and birds (e.g. Hart and Roy, 1966; Berger *et al.*, 1970; Butler and Woakes, 1980; Carpenter, 1985). Breathing and stepping frequencies in trotting and galloping are 1:1 in quadrupedal mammals (Bramble and Carrier, 1983). In many birds wing beat and breathing frequency are also synchronized 1:1 (e.g. Hart and Roy, 1966;

Butler and Woakes, 1980). This is also so for the bat *Pteropus polio-cephalus* (Carpenter, 1985). For birds of high wing loading (ratio of wing area to body weight) values may be considerably higher than a direct 1:1 correspondence. Some ducks, quail and pheasants show values of as high as 5:1 (Berger *et al.*, 1970). It is not yet known whether phase relationships between the respiratory and locomotor systems of animals are associated with any overall energetic advantage. However, whilst the possibility exists, doubt must be cast on the validity of 'static prepara-tions' to assess the respiratory energetics and efficiency of free-ranging animals. These issues are discussed further in Chapter 8.

Similar issues to those outlined above for the possible energetic sig-nificance of respiration-locomotion coupling apply to the cardiovascular system. It is also likely that muscular activity during exercise influences circulatory costs. It is known that skeletal muscle activity improves ve-nous blood return to the heart (Astrand and Rodahl, 1977). In addition, overall arterial anatomy and body shape have an important influence on aortic pressure wave patterns in mammals (e.g. Avolio *et al.*, 1976, 1984). However, the energetic consequences of the distribution and ac-tion of muscle in working animals have not yet been worked out.

2.2   CONCLUDING COMMENTS

Selection can not act on efficiency directly because efficiency parame-ters are derived. Assuming that any biological process that minimizes energy expenditure will be selected for, it follows that high efficiency is adaptive. There are many instances where selection would seem to have acted strongly in maximizing the efficiency of a particular aspect of morphology, physiology, or behaviour. Such circumstances are often associated with high efficiency values and evolutionary convergence with the adaptation concerned. The independent, evolutionary convergence on the highly efficient tunniform swimming mode in fast, pelagic elasmo-branchs, teleosts, and cetaceans is a good example. However, many situ-ations are less clear, involving simultaneous selection for more than one feature or process. Particular structures may perform 'sub-optimally' because of design and/or operational compromises.

Priede (1985) suggests that energetic efficiency is poorly correlated with evolutionary success. Further, critics of maximization arguments contend that empirical work is needed to establish what, if anything, is being maximized by natural selection. Gould and Lewontin (1979) advance the concept of analyzing organisms as integrated wholes. Discussion of efficiency here treats various efficiencies separately for the sake of convenience and simplicity. This is not meant to imply that an integrative approach is not necessary or worthwhile. On the contrary, although difficult, such an approach would be rewarding.

It is assumed here that maximization of the efficiency of energy transformations will result in increased biomass productivity because important life history characteristics (time to first breeding, survivorship, and fecundity) are assumed to be positively correlated with improvements in efficiency. This may not always be so however. Calow (1985) points out that growth rates in fish can be enhanced by hormonal manipulation, implying that growth is below maximum physiological capacity under normal conditions. This may point to ecological rather than physiological constraints on growth. For example, Calow (1985) suggests that fish that are larger than others in their 'size-class' might be more conspicuous to predators. However, in this context it is worth noting that the efficiency of a given process may be maximal, even though the rate of the process is not.

2.3 SUMMARY

1. Understanding the formalisms and assumptions underlying efficiency expressions is important because numerical values reflect them. Also, definitions and their context must be clear and emphasize whether a given value refers to a stage of a process, an overall process, or a series of processes.

2. Common expectations can arise concerning efficiency values. Care must be taken in assuming values based on indirect sources. A large data base drawn from a variety of approaches may not be sufficient to establish a broad general rule.

3. Interpreting the magnitude of numerical values may be straightforward for high values. A single, optimal solution for a given aspect of morphology, physiology or behaviour probably reflects strong selective pressure for maximizing performance. Such circumstances are commonly associated with convergent evolution. Low values can be more difficult, rather than reflecting poor performance per se, they may imply physical and/or biological constraints on a system. Low efficiency of a particular process may reflect a number of different selective pressures for more than one trait. Ideally, the limits and balance of integrated systems

should be known and considered when interpreting the efficiency of a particular component. Unfortunately, this will rarely be possible.

4. Reconciling physiological measurements of locomotor economy (cost of transport) with biomechanical determinations of efficiency is currently problematic for terrestrial, aquatic and aerial animals. Currently, there is no explanation at the biomechanical and functional morphological level as to why physiologically determined cost of transport values are not sensitive to level of organization, functional design, and kinematic pattern.

5. Information on the energetics and efficiency of animal performance gained in the laboratory may not be directly relevant to field situations where animals are free to select movement styles, activity levels, and duration. Maintained constant velocity movement patterns are probably rare in nature. Predictions of energy efficient movement patterns based on mathematical models are supported by field observations of locomotor behavior.

2.4   ACKNOWLEDGEMENT

I would like to thank the Natural Sciences and Engineering Research Council of Canada for financial support. This work was completed whilst the author was a Killam University Research Fellow at the University of British Columbia.

2.5   REFERENCES

Agostini, E. and Mead, J. (1964). Statics of the respiratory system. In: *Handbook of Physiology.* Section 3, Respiration. Vol. 1, pp. 387–409. Eds. W.O. Fenn and H. Rahn. American Physiological Society: Washington, D.C.

Asmussen, E. (1967). Exercise and regulation of ventilation. *Circ. Res.*, **20**:1–132.

Astrand, P.O. and Rodahl, K. (1977). *Textbook of Work Physiology.* 2nd ed. McGraw-Hill, New York.

Atkinson, D.E. (1977). *Cellular Energy Metabolism and its Regulation.* Academic Press, New York.

Avolio, A.P., Bason, P.T., Gow, B.S., Mang, K. and O'Rourke, M.F. (1976). A comparative study of pulsatile arterial hemodynamics in rabbits and guinea pigs. *Am. J. Physiol.* **230**: 868–75.

Avolio, A.P., Nichols, W.W. and O'Rourke, M.F. (1984). Exaggerated wave reflection in kangaroo stimulates arterial counterpulsation. *Am. J. Physiol.* **246**: R267-R270.

Beamish, F.W.H. (1978). Swimming capacity. In: *Fish Physiology*, vol. 7, W. S. Hoar and D.J. Randall eds. pp.101–87. Academic Press: New York.

Berger, M. (1985). Sauerstoffverbrauch von kolibris (*Colibre coruscans* und *C. thalassinus*) beim horizontalflug. Biona, Report 3, pp. 307–14. W. Nachtigall ed. *Akad. Wiss. Mainz.* G. Fischer, Stuttgart, New York.

Berger, J , Hart, J.S. and Roy, O.Z. (1970). Respiration, oxygen consumption and heart rate in some birds during rest and flight. *Z. Vergl. Physiol.* **66**: 201–14.

Blake, R.W. (1983). *Fish Locomotion.* Cambridge University Press, Cambridge.

Blake, R.W. (1986). Hydrodynamics of swimming in the water boatman, *Cenocorixa bifida. Can. J. Zool.* **64**: 1606–13.

Bramble, D.M. and Carrier, D.R. (1983). Running and breathing in mammals. *Science* **219**: 251–6.

Butler, P.J. and Woakes, A.J. (1980). Heart rate, respiratory frequency and wing beat frequency of free flying barnacle geese *Branta leucopsis. J. exp. Biol.* **85**: 213–26.

Calow, P. (1985). Adaptive aspects of energy allocation. In: *Fish Energetics.* pp. 13–31. Eds. P. Tytler and P. Calow. Croom Helm, London and Sydney.

Carpenter, R.E. (1985). Flight physiology of flying foxes, *Pteropus poliocephalus. J. exp. Biol.* **114**: 619–47.

Cavagna, G.A., Heglund, N.C. and Taylor, C.R. (1977). Mechanical work in terrestrial locomotion: two basic mechanisms for minimizing energy expenditure. *Am. J. Physiol.* **233**: 243–61.

Clark, R.B. (1964). *Dynamics in Metazoan Evolution.* Oxford University Press, Oxford.

Cornish-bowden, A. (1983). Metabolic efficiency: is it a useful concept? *Biochem. Soc. Trans.* **11**. 44–5.

Ellington, C.P. (1984). The aerodynamics of hovering insect flight VI. Lift and power requirements. *Phil. Trans. R. Soc. Lond. B* **305**: 145–81.

Fedak, M.A., Heglund, N.C. and Taylor, C.R. (1982). Energetics and mechanics of terrestrial locomotion II: kinetic energy changes of the limbs and body as a function of speed and body size in birds and mammals. *J. exp. Biol.* **79**: 23–40.

Gaeser, G.A. and Brooks, G.A. (1975). Muscular efficiency during steady state exercise: effects of speed and work rate. *J. Appl. Physiol.* **38**: 1132–9.

Gould, S.J. and Lewontin, R.C. (1979). The spandrels of San Marco and the Panglossian paradigm: A critique of the adaptationist programme. *Proc. R. Soc. Lond. B* **205**: 581–98.

Gray, J. and Hancock, G.J. (1955). The propulsion of sea-urchin spermatozoa. *J. exp. Biol.* **32**: 802–14.

Hart, J.S. and Roy, O.Z. (1966). Respiratory and cardiac responses to flight in pigeons. *Physiol. Zool.* **39**: 291–306.

Heglund, N.C., Cavagna, G.A. and Taylor, C.R. (1982a). Energetics and mechanics of terrestrial locomotion III. Energy changes of the centre of mass as a function of body size in birds and mammals. *J. exp. Biol.* **97**: 41–56.

Heglund, N.C., Fedak, M.A., Taylor, C.R., and Cavagna, G.A. (1982b). Energetics and mechanics of terrestrial locomotion IV. Total mechanical energy changes as a function of speed and body size in birds and mammals. *J. exp. Biol.* **97**: 57–66.

Herreid, C.F. (1981). Energetics of pedestrian arthropods. In: *Locomotion and Energetics in Arthropods.* C.F. Herreid and C.R. Fourtner eds. pp. 491–526. Plenum Press, New York and London.

Hill, A.V. (1964). The effect of load on the heat of shortening of muscle. *Proc. Roy. Soc. Lond. B.* **159**: 297–318.

Kushmerick, M.J. and Davies, R.E. (1969). The chemical energetics of muscle contraction II. The chemistry, efficiency and power of maximally working sartorius muscles. *Proc. R. Soc. Lond. B* **174**: 315–54.

Lighthill, M.J. (1969). Hydromechanics of aquatic animal propulsion: a survey. *Ann. Rev. Fluid Mech.* **1**: 413–46.

Lighthill, M.J. (1970). Aquatic animal propulsion of high hydromechanical efficiency. *J. Fluid Mech.* **44**: 265–301.

Lighthill, M.J. (1975). *Mathematical Biofluiddynamics.* Society for Industrial and Applied Mathematics, Philadelphia.

Lissaman, P.B.S. and Shollenberger, C.A. (1970). Formation flight of birds. *Science* **168**: 1003–5.

Medawar, P.B. and Medawar, J.S. (1983). *Aristotle to Zoos.* Harvard University Press. Cambridge, Massachusetts.

Milsom, W.K. and Vitalis, T.Z. (1984). Pulmonary mechanics and the work of breathing in the lizard, *Gekko gecko. J. exp. Biol.* **113**: 187–202.

Otis, A.B. (1954). The work of breathing. *Physiol. Rev.* **34**: 449–58.

Otis, A.B., Fenn, W.O. and Rahn, H. (1950). Mechanics of breathing in man. *J. Appl. Physiol.* **2**: 592–607.

Pennycuick, C.J. (1972). Soaring behaviour and performance of some East African birds, observed from a motor-glider. *Ibis* **114**:178–218.

Perry, S.F. and Duncker, H.R. (1980). Interrelationship of static mechanical factors and anatomical structure in long evolution. *J. Comp. Physiol.* **138**: 321–34.

Peters, R.H. (1983). *The Ecological Implications of Body Size.* Cambridge University Press. Cambridge.

Priede, I.G. (1985). Metabolic scope in fishes. In: *Fish Energetics,* pp. 33–64. P. Tytler and P. Calow eds. Croom Helm, London and Sydney.

Pugh, L.G.C.E. (1974). The relation of oxygen intake and speed in competition cycling and comparative observations on the bicycle ergometer. *J. Physiol.* **241**: 795–808.

Rayner, J.M.V. (1977). The intermittent flight of birds. In: *Scale Effects in Animal Locomotion,* pp. 437–43 T.J. Pedley ed. Academic Press, London.

Schmidt-Nielsen, K. (1971). Locomotion: energy cost of swimming, flying and running. *Science* **177**: 222–6.

Schmidt-Nielsen, K. (1984). *Scaling: Why is Animal Size So Important?* Cambridge University Press, Cambridge.

Spells, K.E. (1969). Comparative studies in lung mechanics based on a survey of literature data. *Resp. Physiol.* **8**: 37–57.

Stucki, J.W. (1980). The optimal efficiency and the economic degrees of coupling of oxidative phosphorylation. *Eur. J. Biochem.* **109**: 269–83.

Taylor, G.I. (1952). Analysis of the swimming of long narrow animals. *Proc. R. Soc. Lond. A* **214**: 158–83.

Taylor, C.R., Heglund, N.C. and Maloiy, G.M.O. (1982). Energetics and mechanics of terrestrial locomotion I. Metabolic energy consumption as a function of speed and body size in birds and mammals. *J. exp. Biol.* **97**: 1–21.

Tucker, V.A. (1970). Energetic cost of locomotion in animals. *Comp. Biochem. Physiol.* **34**: 841–6.

Watt, W. (1986). Power and efficiency as indices of fitness in metabolic organization. *Am. Nat.* **127**: 629–53.

Webb, P.W. (1975). Hydrodynamics and energetics of fish propulsion. *Bull. Fish. Res. Bd. Can.* **190**: 1–158.

Webb, P.W., Kostecki, P.T. and Stevens, E.D. (1984). The effect of size and swimming speed on locomotor kinematics of rainbow trout. *J. exp. Biol.* **109**: 77–95.

Weis-Fogh, T. (1960). A rubber-like protein in insect cuticle. *J. exp. Biol.* **37**: 889–907.

Weis-Fogh, T. (1972). Energetics of hovering in hummingbirds and *Drosophila*. *J. exp. Biol.* **56**: 79–104.

3

Adapting skeletal muscle to be efficient

C. J. PENNYCUICK

3.1 INTRODUCTION

The efficiency with which a muscle converts chemical free energy into mechanical work is adaptively important in prolonged aerobic locomotion, for example in migration. Cruising locomotion is characterized by cyclic contraction at a well defined frequency, in which each muscle actively shortens, and is then passively stretched back to its original length. The muscle does a certain amount of work (the 'cycle work') in each cycle (Fig. 3.1). The average rate of doing work, that is the mechanical power output, is found by multiplying the cycle work by the contraction frequency.

Since the limbs of vertebrates and arthropods are actuated by muscles directly, without the intervention of any form of reduction drive, the frequency of contraction is necessarily the same as the stepping frequency for a terrestrial animal, or the frequency at which a flying animal beats its wings, or a swimming one beats its tail or flippers. The frequency at which any particular animal oscillates its limbs is determined by mechanical considerations, the general nature of which was discussed by Hill (1950). 'Natural' frequencies of limb oscillation in terrestrial locomotion have been considered by Alexander (1976, 1980), and for flying animals by Pennycuick (1975a, 1990). For present purposes, it is sufficient to note first, that the range of frequencies available to a particular animal in cruising locomotion is narrow, and second, that there is a strong trend for larger animals to oscillate their limbs at lower frequencies than smaller ones of similar general type. This latter trend is plainly visible to the naked eye. Walking children step at higher frequencies than their

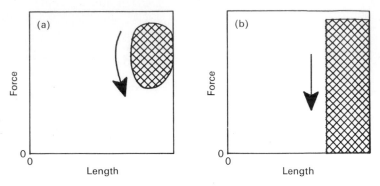

Fig. 3.1. If the force exerted by a muscle is plotted against the length as it performs regular cycles of contraction and extension, the point traces out a closed loop, once for each cycle. If the point travels anticlockwise as shown, the force is higher in shortening than in lengthening, and the muscle does work. The area enclosed by the loop (hatched) represents the amount of work done in each cycle. The diagram looks similar if stress instead of force is plotted on the y-axis, and strain instead of length on the x-axis, but in this case the area of the loop represents volume-specific work. Two types of loop are shown - in (a) there is a large amount of residual force during lengthening, as in insect flight muscles. In the simplified theory of vertebrate locomotor muscles it is assumed that the force is constant during shortening, and drops abruptly to zero for lengthening, as in (b).

parents, large whales beat their tails at lower frequencies than dolphins, and hummingbirds have no choice but to beat their wings at higher frequencies than pelicans. The frequency at which a muscle contracts is imposed upon it by the mechanical properties of the load to which it is attached.

3.2   POWER OUTPUT AND EFFICIENCY OF LOCOMOTOR MUSCLES

3.2.1   Specific power output of muscle

Rather than considering the absolute amount of cycle work produced by a particular muscle, a more general formulation is to consider either the volume-specific or the mass-specific work, that is the work done by unit volume, or mass, of muscle respectively. It was shown by Pennycuick and Rezende (1984) that the volume specific work (Q_v) for a muscle performing a cycle like that of Fig. 3.1b may be found by multiplying the stress (T) during shortening by the active strain (L). The stress is the force exerted per unit cross-sectional area of the muscle, and the active strain is the shortening distance, divided by the extended length of the muscle. The volume-specific power output (P_v) is then found by multiplying the volume-specific work by the contraction frequency (f):

$$P_v = Q_v f = TLf \tag{3.1}$$

Confining attention for the moment to the contractile proteins themselves, the stress they are able to exert whilst shortening is closely related to the tension exerted by each myosin filament, and therefore not likely to vary much from one muscle to another of the same general type. Similarly, the strain is an expression of the minimum and maximum overlap between myosin and actin filaments, and is determined by the geometry of the sarcomere. The third variable in Equation (3.1), the contraction frequency, varies widely in different animals. There is a strong trend for larger animals to have lower contraction frequencies, and therefore lower volume-specific power outputs as well. In aerobic muscles, volume-specific power output is not proportional to contraction frequency, as Equation (3.1) might suggest, because muscles operating at higher frequencies need more mitochondria to support the higher volume-specific power output of the myofibrils. At very high frequencies, as in flying insects and hummingbirds, the volume-specific power approaches a limiting value that is determined more by the capabilities of the mitochondria than by the mechanical properties of the myofibrils (Pennycuick and Rezende, 1984).

3.2.2   *Efficiency related to contraction frequency*

In cyclic contraction, a muscle shortens through some fixed amount of strain, in approximately half the time required for the complete cycle of shortening and lengthening. Therefore the contraction frequency is closely related to the strain rate (V) during shortening, defined as the speed of shortening divided by the extended length of the muscle. This allows efficiency to be related to contraction frequency, using the concepts introduced by Hill (1938), in his famous study of the production of heat and work by muscles contracting against a constant force. Hill studied rapid mechanical and thermal events, in which the muscle goes through a cycle of changes, and returns to its original state. In this restricted context, it can be assumed that the free energy supplied by dephosphorylation of ATP is converted into either work or heat, so that the efficiency can be represented as

$$Efficiency = Work/(Work + Heat). \qquad (3.2)$$

Equation (3.2) does not necessarily hold if the 'system' under scrutiny also includes the oxidative reactions that regenerate the ATP, because in this case some of the free energy consumed may become 'bound energy', representing a decrease of entropy of the reaction products as compared to the reactants. In this broader context, Gnaiger (1989) shows that Equation (3.2) is nevertheless a good approximation in aerobic, though not in anaerobic metabolism.

No work is done if the muscle contracts against zero force, although fuel energy is consumed and heat is produced, so in this case the efficiency is zero. The efficiency is also zero if the force is high enough to prevent the muscle from shortening (isometric contraction). In this case also, heat is produced but no work is done. At intermediate values of the force, work as well as heat is produced, and the efficiency is positive. The curve of Fig. 3.2b shows efficiency plotted against the ratio of the stress to the isometric stress. The curve shows a broad maximum, and its shape is the same as it would be if force were used as the ordinate.

3.2.3   Efficiency from Huxley's equations

The curves of Fig. 3.2 were calculated from the quantitative sliding filament model of muscle developed by Huxley (1957), which successfully accounted for Hill's (1938) observations. McMahon (1984), reviewing the predictions of Huxley's model in detail, noted that it was also very successful in accounting for later evidence, not known to Huxley at the time the model was formulated. McMahon presented the equations in a form from which the efficiency can easily be derived. In McMahon's notation, T is the stress exerted by the fully activated muscle, and V is the strain rate, as above. Efficiency can be found by comparing the model's estimates of the mechanical power output, and of the rate at which energy is supplied by splitting ATP. Both rates now refer to a muscle that is shortening at a constant strain rate, and are not averaged over a number of cycles of contraction and relaxation, as above. The product TV is the instantaneous volume-specific mechanical power output, that is the rate of doing work, per unit volume of muscle, while the muscle is actually shortening. E is the volume-specific rate at which energy is supplied by splitting ATP (again, during shortening), and the efficiency η is

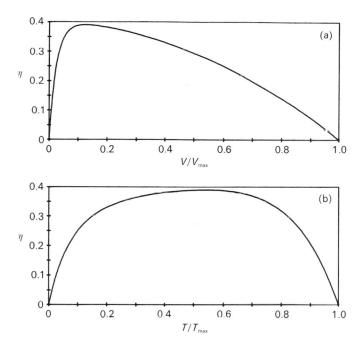

Fig. 3.2. Efficiency plotted against (a) normalized strain rate and (b) normalized stress.

$$\eta = TV/E. \qquad (3.3)$$

This formulation does not require any assumption as to whether the remaining energy is converted into heat or bound energy. McMahon's E, being based on the rate of splitting ATP molecules, does not take account of losses that occur in the creation of ATP by oxidizing fat or carbohydrate. The estimate of efficiency is therefore a theoretical maximum, and is higher than the estimates obtained by physiological experiments in which an increment of oxygen consumption is compared with an increment of mechanical power output. However the present paper is concerned primarily with the way in which the efficiency varies with the stress and the strain rate, rather with its absolute value. The stress (T) is given by McMahon's Equation 4.32 as

$$T = (msw/2l)[f_1/(f_1 + g_1)]$$
$$\{1 - (V/\Theta)[1 - \exp(-\Theta/V)][1 + 1/2((f_1 + g_1)/g_2)^2(V/\Theta)]\} \quad (3.4)$$

The volume-specific rate of energy consumption (E) is given by McMahon's Equation 4.37 as

$$E = (meh/2l)[f_1/(f_1 + g_1)]g_1 + f_1(V/\Theta)[1 - \exp(-\Theta/V)] \qquad (3.5)$$

When T is divided by E to obtain the efficiency, two of the variables cancel out, that is m, the number of cross-bridges per unit volume of muscle, and l, the distance between adjacent attachment sites on the actin filaments. Other variables can be replaced by numerical ratios. f_1 and g_1 are the rate constants (with dimensions of inverse time) for attachment and detachment of cross bridges in the forward direction, and are related to each other so that the ratio $f_1/(f_1 + g_1) = 13/16$; g_2 is the rate constant for detachment in the reverse direction, and the ratio $g_2/(f_1 + g_1) = 3.919$. Θ is a dimensionless ratio that relates the maximum value of V (against zero stress) to the forward rate constants, and to the extended length of a sarcomere (s), and maximum deflection of a cross bridge (h):

$$\Theta = V_{max}/4 = (h/s)(f_1 + g_1) \qquad (3.6)$$

The remaining variables in the above equations are the maximum work that can be done in one cross bridge cycle (w) and the energy liberated by splitting ATP in one cross bridge cycle (e). These have the ratio $w/e = 3/4$. After simplifying in accordance with these values, the efficiency (η) can be expressed in terms of a single variable, the dimensionless ratio V/V_{max}.

$$\eta = TV/E =$$
$$(48V/13V_{max})1 - (4V/V_{max})[1 - \exp(-V_{max}/V)][1 + (0.13V/V_{max})]/$$
$$\{(3/13) + (4V/V_{max})[1 - \exp(-V_{max}/V)]\} \qquad (3.7)$$

V/V_{max} varies from zero in an isometric contraction, to 1 when the muscle shortens against zero force. The curve of η versus V/V_{max} is shown in Fig. 3.2a. The efficiency rises steeply to a maximum value of just over 0.39 when the ratio V/V_{max} is about 0.13, and thereafter declines progressively to zero at $V/V_{max}=1$.

3.2.4   Adaptation of a muscle for maximum efficiency

If efficiency is an important adaptive consideration, then Fig. 3.2a shows that the contraction frequency must be such as to keep the strain rate at approximately 13% of its maximum value. Efficiency drops quickly if the strain rate is allowed to depart from this value, especially on the low side of the maximum. The ratio V/V_{max} must remain within quite narrow limits. The value of V (the strain rate) is determined by the size and geometry of the animal, and the nature of the load against which the muscles have to work. That is, the value of V at which the muscles must shorten is imposed on the animal V_{max} that has to be adjusted. This is effected by adjusting the rate constants for attachment and detachment of myosin cross bridges (Equation (3.6) above). The value of V_{max} must be adjusted to be about 7.7 times the imposed value of V, to achieve $V/V_{max} = 0.13$. The notion of an 'operating frequency' as used by Pennycuick and Rezende (1984) refers to the contraction frequency at which a particular muscle is adapted to operate efficiently, by virtue of its particular value of V_{max}. V_{max} (called the 'intrinsic speed' by Hill, 1938) is the property by which a particular muscle is characterized as 'slow' or 'fast'. Larger animals must have slower locomotor muscles than smaller animals of similar geometry. In steady locomotion, the contraction frequency also determines the mechanical power output of the muscles (Equation (3.1)), subject to the effect of mitochondria in aerobic muscles as noted above. The power output in turn determines the rate at which the enzyme systems are required to supply energy to the contractile proteins in the form of ATP.

It has sometimes been suggested that the mechanical power output of muscles is limited by the capacity of the enzyme systems to supply energy, but this is not so. The mechanical power output is determined by the contraction frequency, which is in turn determined by mechanical constraints resulting from the size and geometry of the animal. This determines the value to which V_{max} must be set, which is equivalent to specifying a required value for the cycling frequency of the myosin cross bridges. The rate at which ATP is required is directly related to this cycling frequency, and the enzyme systems have to be adapted to supply it at the rate so determined. There is no advantage in having enzyme systems that can supply ATP at a higher rate, since any additional energy supplied could only be turned into heat, not into work. The power output of locomotor muscles is limited in the first instance by mechanical considerations. The enzyme systems are adapted secondarily, to supply energy at a rate at which the contractile proteins are able to use it.

3.2.5 Flight at reduced power

A migrating bird has to fly near its maximum range speed if it is to maximize the distance flown in consuming its load of fuel. For reasons explained by Pennycuick (1975a), the progressive reduction of the bird's weight, as fuel is consumed, results in a corresponding reduction of the maximum range speed, and a massive reduction in the power required to maintain this speed. To remain efficient aerodynamically, the bird must exert progressively less power with its flight muscles. One could imagine that it might do this by reducing either the frequency or the amplitude of flapping. However, either expedient would reduce the strain rate during shortening, and if the muscle were operating near peak efficiency at the beginning of the flight, these changes would result in reduced efficiency (Fig. 3.2a). This method of reducing power can be seen in short-term maneuvers such as brief descents, but is probably never used by birds in prolonged horizontal flight. Instead, power is reduced by flapping intermittently. Passerines and some other small birds reduce power by 'bounding', in which a flapping phase alternates with a ballistic phase. In the flapping phase, the bird exerts more power than is needed to maintain level flight, so that the flight path curves upwards. Then, the wings are folded up, resulting in a short, downward curving, ballistic trajectory. Most medium-sized and large birds reduce power by 'flap-gliding', in which a flapping phase alternates with a gliding phase. This method permits level flight, but the bird accelerates during the flapping phase and slows down during the gliding phase. In either case, the mechanical conditions during the flapping phase can remain unchanged (and presumably near peak efficiency), while the average power is reduced. Rayner (1985) argues that bounding may be energetically advantageous for aerodynamic reasons alone. Whether or not this is so, considerations of muscle efficiency would suggest that some form of intermittent flight is preferable to other methods of reducing power. In a passerine on a long migratory flight, one would expect to see the bird flapping continuously, or nearly so, at the beginning of the flight, with the ballistic phase occupying a progressively increasing proportion of the time, as fuel is used up.

3.2.6   Cost of maintaining tension

All muscles can produce tension, but not all are adapted to produce work. 'Tonic' or 'postural' muscles are adapted to maintain tension without shortening. A muscle doing this consumes energy, but does no work, so that its efficiency as defined above is zero. A more relevant measure of muscle performance in this case is the energetic cost of maintaining force, that is the ratio of the power (rate of consuming chemical energy) to the force. Since power is force times speed, the ratio of power to force has the dimensions of speed (LT^{-1}). It is not difficult to guess that it is closely related to the maximum speed of shortening of the muscle, a connection which has been known in general terms to physiologists for many years (Johnston, 1985). As usual it is more convenient to replace force and power by variables that do not depend on the size or shape of the muscle. In terms of McMahon's (1984) notation, T is the stress, or force exerted per unit cross-sectional area, and E is the rate at which energy is released by splitting ATP, per unit volume of muscle. That is, E is a volume-specific power. Dividing a volume-specific power by a stress yields a quantity with the dimensions of inverse time (T^{-1}). McMahon's equations 4.33 and 4.38 give expressions for T_0 and E_0, defined as the stress and volume-specific power when the speed of shortening is zero. Dividing E_0 by T_0 gives the energetic cost of maintaining the maximum isometric tension, as follows:

$$E_0 = (mehg_1/2l)[f_1/(f_1 + g_1)]T_0 = (msw/2l)[f_1/(f_1 + g_1)] \qquad (3.8)$$

The variables have the same meanings as above, under the discussion of efficiency. Dividing E_0 by T_0, and simplifying in the same manner as above:

$$E_0/T_0 = V_{max}/16 \qquad (3.9)$$

It will be remembered that in McMahon's notation, V_{max} is the maximum value for the strain rate (not the speed of shortening). It has the appropriate dimensions of inverse time. V_{max} may be measured for a particular muscle, by determining the speed at which the muscle can shorten against zero load, and dividing this by the extended length of the muscle. This value (in s^{-1}), if divided by 16, and multiplied by a stress (in pascals) will yield the power (in watts) needed to maintain this stress, in terms of the required rate of splitting ATP molecules. As noted above in the discussion of efficiency, the power in terms of the rate of consumption of fuel energy will be higher, on account of energy losses in the synthesis of ATP.

3.3   CONCLUSION

Widely different mechanical demands are made on locomotor muscles, depending on the type of locomotion for which they are used, and above all on the size of the animal which they propel. The adaptation by which a particular muscle is matched to the mechanical demands made upon it consists in adjusting its 'intrinsic speed' (maximum strain rate) to suit the strain rate required in locomotion. The locomotor muscles of a large animal must be slower than those of a similar but smaller one, and the muscles of a growing animal must be progressively slowed down as its size increases. Slower muscles also produce less specific power output than faster ones, and this effect cannot be circumvented by biochemical adaptations. The energetic cost of maintaining a steady force is also determined by the maximum strain rate, but there is no optimum value in this case. The V_{max} of a postural muscle has to be high enough to allow necessary adjustments of posture, but subject to that, the slower the muscle, the lower the cost. If no adjustment of length is necessary, then the muscle can be replaced by a ligament, which maintains tension without any expenditure of energy.

3.4 REFERENCES

Alexander, R.McN. (1976). Mechanics of bipedal locomotion. In: *Perspectives in Experimental Biology*, ed. P. Spencer Davies, pp. 493–504. Oxford, Pergamon.

Alexander, R.McN. (1980). Optimum walking techniques for quadrupeds and bipeds. *Journal of Zoology*, **192**, 97–117.

Gnaiger, E. (1989). Physiological calorimetry: heat flux, metabolic flux, entropy and power. *Thermochimica Acta* (in press).

Hill, A.V. (1938). The heat of shortening and the dynamic constants of muscle. *Proceedings of the Royal Society B*, **126**, 136–95.

Hill, A.V. (1950). The dimensions of animals and their muscular dynamics. *Science Progress*, **38**, 209–30.

Huxley, A.F. (1957). Muscle structure and theories of contraction. *Progress in Biophysics and Biophysical Chemistry*, **7**, 255–318.

Johnston, I.A. (1985). Sustained force development: specializations and variation among the vertebrates. *Journal of Experimental Biology*, **115**, 239–251.

McMahon, T.A. (1984). *Muscles, reflexes and locomotion*. Princeton, Princeton University Press.

Pennycuick, C.J. (1975a). Mechanics of flight. In: *Avian Biology*, vol. 5, ed. D.S. Farner & J.R. King, pp. 1–75. New York, Academic Press.

Pennycuick, C.J. (1975b). On the running of the gnu (*Connochaetes taurinus*) and other animals. *Journal of Experimental Biology*, **63**, 775–99.

Pennycuick, C.J. (1990). Predicting wingbeat frequency and wavelength of birds. *Journal of Experimental Biology*, in press.

Pennycuick, C.J. & Rezende, M.A. (1984). The specific power output of aerobic muscle, related to the power density of mitochondria. *Journal of Experimental Biology*, **108**, 377–92.

Rayner, J.M.V. (1985). Bounding and undulating flight in birds. *Journal of Theoretical Biology*, **117**, 47–77.

4

Efficiency and other criteria for evaluating the quality of structural biomaterials

J. M. GOSLINE

4.1 INTRODUCTION

It is often assumed that the structural materials and mechanical devices found in animal skeletons represent perfect or near perfect solutions to the multitude of functional requirements that organisms encounter in their adaptation to ecological niches. That is, it is assumed that evolution has produced optimal designs in the construction of its skeletal systems. It is not clear, however, that this optimistic viewpoint is justified. Evolutionary biologists generally reject the concept of DESIGN as being 'teleological', because it implies that evolution is directed to a set of specific goals. Further, each structure and mechanism likely represent a compromise between a multitude of conflicting factors that together determine ecological fitness, rather than a clear optimum reflecting a single functional attribute. It is therefore reasonable to ask if design is a useful concept when considering the evolution of skeletal structures.

The word design has several meanings, and this is the cause of much confusion. Design as a noun is simply a statement of the relationship between the structure and the function of an object. For example, a bird or an aircraft wing has a shape that can be associated with its ability to generate aerodynamic lift with minimal drag, and this shape can then be regarded as a feature of the wing design. Alternately, the molecular and microscopic structure of collagen fibres can be correlated with the skeletal function of the tendons and ligaments that are fashioned from them. Again, the collagen fibre is a design because there appears to be a sensible relationship between structure and function. This definition of Design is relatively neutral because it does not imply quality or optimality in construction. There are good designs in which the relationship between structure and function is strong, and poor designs in which this relationship is weak. It is this definition of design that will be used in this paper. Design can also be a verb, which represents the process through which functioning systems are conceived, produced, evaluated and improved, until some functional end product is constructed. In this process structures (i.e. designs) are produced, evaluated and refined, by natural selection in the case of evolution. This process might be thought to lead to the structural perfection alluded to above. The nature of this design process in living systems (i.e. evolution) is an extremely contentious issue and one which will not be resolved in the few pages of this essay. Why then should we consider the philosophy of design in a discussion of efficiency as criterion for evaluating the quality of biomaterials?

The answer is that over the past several hundred years the science of engineering has developed a rational approach to evaluating the quality of mechanical structures. In addition, in the act of creating a vast array of structures, engineers have actually carried out the process of design, from the conception of functional goals, to the evaluation of the test designs, to the balancing of the various economic, geographical and sociological factors which determine the success of a design, to the creation of the final product. That is, there are strong parallels between the science and practice of human engineering and the evolution of organismal engineering. This Chapter attempts to apply engineering theory to the assessment of quality in the design of biomaterials. If we succeed in this initial attempt, then perhaps the study of organismal engineering, or biomechanics, may provide useful insights into the process of evolution.

4.1.1 *The efficiency of biomaterials*

The concept of efficiency is often taken to represent the balance between some functional output relative to the input required to achieve this output. As such, efficiency may provide a useful criterion to evaluate the quality of biomaterials, and it is tempting to speculate that the process of evolution is directed by the selection of systems which achieve increases in efficiency. However, this definition of efficiency is so vague as to make it useless in any rational discussion of either the process of evolution or the design of biomaterials. I will begin, therefore, by using the thermodynamic definition of efficiency, which relates specifically to the effectiveness of energy conversions in chemical or physical systems. That is, efficiency is the ratio of the energy or power output by a system to the energy or power input to that system. Such an efficiency is a unit-less ratio which, according to the laws of thermodynamics must fall between the values of zero and one. How useful is this type of ratio in evaluating the quality of biomaterials design?

The only direct application of thermodynamic efficiency to structural materials is in the concept of elastic efficiency or resilience, and for materials that function in energy exchange mechanisms resilience may provide a useful criterion. It is unlikely, however, that resilience is important in all skeletal mechanisms because many aspects of mechanical support do not involve energy exchanges, and even in systems where energetic exchanges are important, it is not necessary that elastic efficiency be maximized for optimal function of the system. Some examples should clarify these statements.

Elastic energy storage mechanisms play a major role in reducing the metabolic cost of locomotion in many animals (Alexander, 1988), and high elastic efficiency is likely to be a desirable attribute of the materials used in the construction of such energy storage systems. In insect flight, for example, elastic structures in the wing hinges made from the rubber-like cuticular protein resilin or from rigid cuticle provide springs that virtually eliminate the costs of overcoming the inertial forces associated with high frequency wing oscillations. Once the wing is set in motion, the kinetic energy of the moving wing is absorbed and stored by elastic structures that bring the wing to a halt at the end of its travel, and elastic recoil provides the energy to accelerate the wing in the opposite direction. Thus, muscular effort is required only to overcome the aerodynamic forces that are responsible for lift and thrust (Weis-Fogh, 1972). Similarly, tendons and ligaments in vertebrate skeletons provide elastic structures that store and release energy associated with fluctuations in the velocity and height of running birds and mammals (Alexander, 1984). In both of these situations the resilience of the structural materials will strongly affect the ability of energy storage systems to reduce metabolic costs to the animal.

Figure 4.1. shows a typical load-unload cycle for a sheep plantaris tendon, based on the work of Ker (1981). The specimen was cycled at 11 Hz, a rate of deformation similar to that which occurs in normal locomotion, and resilience can be determined directly from the data (Wainwright *et al.*, 1982). The energy required to stretch the tendon (E_{in}) is equal to the area under the load curve; whereas the energy recovered as elastic recoil (E_{out}) is equal to the area under the unload curve. The energy lost as heat in the cycle, equal to the shaded area enclosed by the load

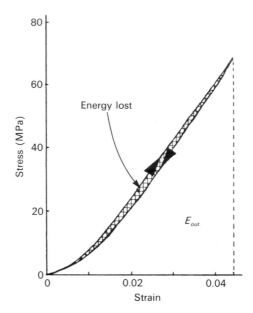

Fig. 4.1. Stress-strain diagram for sheep plantaris tendon undergoing a load-unload cycle. The shaded area, bounded by the load cycle, is proportional to the energy dissipated as heat in the cycle. The area under the unload curve and bounded by the dashed line, labelled E_{out}, is proportional to the energy recovered through elastic recoil. The total energy required to deform the tendon, E_{in}, is equal to the sum of the energy lost and E_{out}. The resilience of the tendon is approximately 93%. This figure is based on the data of Ker (1981).

and unload curves, is very small. Thus, the resilience (R), defined as $R = E_{out}/E_{in}$, is high, at about 0.93 or a 93%. The resilience of resilin and of insect cuticle is about the same magnitude (Jensen & Weis-Fogh, 1962), and since a resilience of this level is high for most polymeric materials, it seems likely that the evolution of vertebrate tendon and of the insect flight system has been directed by the selection of biomaterials with increased resilience.

Elastic efficiency need not, however, be maximized in all systems. Figure 4.2 shows similar load-cycle information for another biomaterial that functions in an energy exchange mechanism. The material is the major ampullate gland silk that makes up the polygonal frame and the radial spokes of the orb-web constructed by the spider *Araneus*. One of the main functions of this material is to absorb the kinetic energy of an insect that flies into the web, and to do this in a manner that aids in the capture of the insect by the spider. The stress-strain diagram shown in

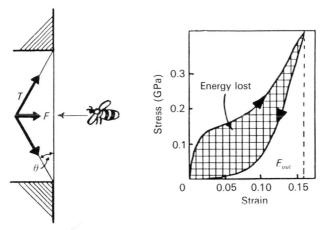

Fig. 4.2. The mechanical properties of the major ampullate gland silk of the spider *Araneus*, based on the work of Denny (1976). The energy lost and the elastic energy recovered in the load cycle are as described in the legend to Fig. 4.1, but in this case the resilience is much lower, at approximately 35%. The diagram to the right shows the normal manner in which flying insects would impact with silk strands in an orb web. The web fiber is viewed from the side, and the insect strikes the web while flying at right angles to the fiber direction. This causes the fiber to deflect, as shown. If the silk had high resilience then the elastic recoil of the system might catapult the prey out of the web. The ratio of the force (F), that acts to resist the forward motion of the insect, to the tension (T) in the silk fiber is proportional to the sine of the deflection angle (Θ) For small deflection angles, the F/T is small; thus, inextensible fibers will break under small loads.

this figure is based on the data of Denny (1976), and it shows a typical load cycle that is quite different from that of vertebrate tendon. Again, the area bounded by the load cycle represents the energy lost as heat in the cycle, but for frame silk this area is quite large. In fact, Denny calculates the resilience to be only about 35%. In this case the material functions by converting kinetic energy into heat, and thus the function requires low elastic efficiency. If the web material behaved like a tendon and had high resilience, then most of the kinetic energy of the insect's movement as would be stored as elastic energy and the subsequent elastic recoil would propel the prey out of the web like a catapult. Clearly, evolution may be directed by the selection of materials that minimize elastic efficiency.

A final possibility is that in some systems materials may function optimally with an intermediate level of elastic efficiency. The elastic arteries of the vertebrate circulatory system provide an interesting example of this. The major vessels of the arterial system contain a rubber-like protein called elastin which allows the arteries to expand and store elastic energy during cardiac systole. The elastic recoil of these vessels maintains the flow of blood to the tissues during cardiac diastole when the heart is refilling. This energy storage system smooths the pulsatile flow of blood from the heart, lowers blood pressures and reduces the work load of the heart. As with vertebrate tendon, one might expect that evolution would have produced arterial elastic tissues that maximize elastic efficiency, and elastin, which makes up as much as 60% of the dry weight of the arterial wall, has a resilience greater than 90% for deformations at cardiac frequencies (Gosline, 1980). The arterial elastic tissue, however, is a complex composite of elastin, collagen fibres and smooth muscle cells, and this tissue has a resilience of about 65% (Wainwright *et al.*, 1982). Why, if evolution leads to increased efficiency, is the resilience of arterial elastic tissues so much lower than the resilience of its major structural component, elastin? The answer lies in the specific design of the arterial system. The cardiovascular system is tuned to take advantage of the propagation, reflection and interference of pressure waves in the arteries, as described in detail in Chapter 9 of this volume. The main feature of this tuning is that pressure waves at the cardiac frequency are reflected from sites in the periphery approximately one quarter wavelength from the heart, and when these reflected waves return to the heart they are approximately out-of-phase with the next pressure wave. The pressure waves combine destructively to decrease blood pressure dramatically, and this reduces the work of the heart. Higher frequency components of the cardiac pressure pulse, however, have shorter wavelengths and when reflected can combine in phase to raise blood pressure at the heart. The relatively low resilience of the elastic tissue allows a reasonable portion of the low frequency pressure waves to be reflected back to the heart, but the high frequency pressure waves are largely attenuated. Thus, the resilience of arterial elastic tissues is a compromise that allows an appropriate balance of elastic recoil and attenuation.

These examples suggest that thermodynamic efficiency can be useful in evaluating the quality of mechanical design in some biomaterials, particularly those materials which function in energy exchange systems, but it is clearly wrong to assume that evolution necessarily leads to materials that maximize efficiency. What about the materials that function to provide support or protection rather than energy exchange mechanisms?

Table 4.1. *Effectiveness can be assessed in ratios of a functional mechanical output divided by the cost or input. This table lists outputs and inputs for effectiveness ratios*

Output	Input
Strength	Cost of synthesis
Toughness	Cost of maintenance
Durability	Cost of evolution
Extensibility	

4.1.2   Effectiveness criteria for biomaterials

Although thermodynamic efficiency is unlikely to provide a universal criterion for evaluating biomaterials, the general concept of an efficiency, the output relative to the cost or input required to achieve this output, is likely to provide a good basis for evaluating the quality of a biomaterial, or for that matter, any functional system created by human or by evolutionary design processes. It will be useful, however, to give such output/input ratios a distinct name and reserve the term efficiency for unit-less ratios of energies. As suggested by Gans and by Lauder (Chapters 1 and 10, respectively), the term effectiveness will be used for such benefit/cost ratios, in which the output (benefit) and input (cost) need not be energies and indeed need not have the same units.

The problem, however, is that the choice of design criteria requires an intimate knowledge of function, but biomechanical systems can have many different functions and many of these functions may not yet even be recognized. Indeed, many criteria can be applied, and the problem, in part, becomes deciding which criteria, are best. Table 4.1 lists a number of outputs and inputs which may apply to the design of biomaterials. The outputs are fairly familiar and are taken directly from engineering. The inputs listed here range from simple physical quantities such as mass or density to complex biological costs. Some costs, such as the cost of synthesis, are quite straightforward, and reflect the fact that an animal must utilize raw materials and energy to synthesize, modify or repair the structural components of its skeleton. A vague concept like the cost of evolution, however, is much more difficult to define. It may in part express the fact that evolution is a historical process,

Table 4.2. *Some functional mechanical attributes and the mechanical properties and units used to quantify these attributes.*

Functional attribute	Mechanical property	Units
Stiffness	Modulus of elasticity	Nm^{-2} (Pa)
Strength	Stess at fracture	Nm^{-2} (Pa)
Toughness	Work of fracture	Jm^{-2}
Durability	Fatigue lifetime	Load cycles
Extensibility	Strain at fracture	no units

and the modification of an existing component is more likely than the development of a novel structure to carry out the same function.

Effectiveness can now be expressed in terms of input/output ratios, such as stiffness per unit cost of synthesis, etc., and if selected appropriately these ratios may actually reflect the attributes that natural selection acts upon in the process of evolution. For example, in a structural system requiring strength, a mutation that increases cross-linking in a biopolymer network may increase the fitness of the organism. Conversely, the incorporation of a rare, essential amino acid may cause an animal to divert limited resources away from other important functional systems and hence reduce fitness. We are left, however, with the problem of how to measure these effectiveness ratios.

4.1.3   Properties

If design is the relationship between structure and function, and our objective is to evaluate the quality of design, we need some means of quantitatively assessing this structure-function relationship. As described by Wainwright (1988), the linkage between structure and function is provided by properties, which by definition are the specific, measurable attributes of a system. In the study of biomaterials we are familiar with a variety of chemical and physical properties that relate to macromolecular structure, including chemical composition, monomer sequence, 3-D crystal structure, molecular orientation, volume fraction of crystalline and non-crystalline phases, etc. In addition, engineers have devised measurable, mechanical properties associated with the functional attributes of structural systems and have established straightforward experimental protocols for measuring these properties. Table 4.2 lists some mechanical properties and their units for the functional attributes identified in Table 4.1. Thus, Modulus of Elasticity is the mechanical property used to measure stiffness, and it has units of Newtons per square meter. It should be noted that these properties were created by engineers to quantify attributes that were recognized to be important in the functioning of man-made systems. It is likely that these attributes are important in biological systems as well and that they will provide good measures of the functional outputs needed for determining effectiveness ratios. On the other hand, living systems may contain novel functional attributes that require new properties in addition to the standard engineering ones.

Quantification of the inputs for effectiveness ratios can be considerably more difficult. With the exception of the simple physical measure of weight or density, the biological costs listed in Table 4.1 range from very difficult to virtually impossible to measure. The energy content of a material can be determined by calorimetry, but this measure is unlikely to provide even a close estimate of the cost of synthesizing that material. The cost of maintaining or repairing a material will be even more difficult to quantify, and the cost of evolution is almost impossible to define, let alone measure. In short, effectiveness ratios based on evaluating mechanical outputs relative to complex biological input costs are virtually unobtainable at this time. We are left, therefore, with ratios based on mechanical output per unit weight, or as normally expressed, specific mechanical properties determined as the mechanical property divided by density. The result of our inability to quantify complex, biological input costs is that we must virtually abandon any attempt to measure effectiveness ratios that might be useful in understanding the processes that direct the evolution of biomaterials. We are left, however, with ratios which may help us understand the quality of mechanical design in biomaterials.

The situation for man-made materials is quite different. Economic data provide estimates of the amount of energy needed to produce various materials, and effectiveness ratios based on energy costs can be very instructive (Gordon, 1978). High quality materials such as titanium or carbon fiber composites require orders of magnitude more energy to provide a given amount of stiffness or strength than ordinary materials such as steel, concrete or wood, and these data go a long way to explain why materials like wood and concrete are so prevalent in static structures like buildings. For moving structures like airplanes the choice of material depends much more on stiffness or strength per unit weight than on economic costs. Thus, density-specific properties are frequently used to evaluate man-made materials. Because most animals are motile, weight considerations also will be important, and it is likely that specific properties will provide insights into the design of biomaterials.

4.1.4   The quality of mechanical design in biomaterials

Armed with this understanding of effectiveness ratios and with a set of mechanical properties to measure functional attributes, we can begin to evaluate some biomaterials. Tables 4.3, 4.4 and 4.5 provide values for density-specific stiffness, strength and toughness for a very short list of man-made and natural structural materials. The data in these tables were taken from Wainwright *et al.* (1982) and Gordon (1978) and have been rounded off to give reasonably accurate values. More detailed information is available in these books and in Vincent & Currey (1980) and Vincent (1982). Toughness is expressed as the energy to break and has been calculated from the area under a tensile stress-strain curve to the point of fracture. The specific properties are expressed as the fundamental property divided by specific gravity (i.e. density relative to that of water), as this keeps the units of the specific property the same as that of the fundamental property. Obviously, the list does not contain all biomaterials; rather it provides some useful examples which will allow us to discuss the selection of mechanical properties that adequately reflect important function(s) of biomaterials. Stiffness, or specific stiffness, as listed in Table 4.3, is the property that is most often given for structural materials in the engineering literature, and this preference arises from the fact that most engineering structures are rigid ones for which stiffness is a crucial property. Vertebrates and arthropods have rigid skeletons, so it seems logical to compare their skeletal materials with engineering materials. When we compare the stiffness or specific stiffness of tendon collagen, the tensile fibers of the vertebrate skeleton, and of bone and cuticle, the rigid materials of vertebrates and insects, with their engineering equivalents, we get the impression that the biological materials are dramatically inferior. For example, Kevlar and carbon fibers are roughly two orders of magnitude stiffer than tendon, and carbon fiber composites are about an order of magnitude stiffer than bone or cuticle. Even steel, with a specific gravity of 7.8, has more than twice the specific stiffness of bone and cuticle. Indeed, bone and cuticle seem roughly equal in stiffness to the rather modest ceramic material, concrete. Does

Table 4.3. *Tensile modulus of elasticity, specific gravity and specific modulus of elasticity data for selected natural and synthetic materials. The data are from Wainwright* et al. *(1982) and Gordon (1978)*

Material	Modulus (GPa)	Sp. gravity	Sp. modulus (GPa)
TENSILE FIBERS			
Tendon (collagen)	1.2	1.3	1
Spider's frame silk	10	1.3	8
Kevlar	130	1.45	90
Carbon fiber	400	2.2	180
RIGID MATERIALS			
Bone	20	2.2	9
Insect cuticle	9	1.2	7
Carbon fiber composite	150	1.8	80
High tensile steel	210	7.8	20
Glass	70	2.5	30
Concrete	20	2.5	8

this mean that biomaterials are low quality designs? This apparent disparity in quality is seen again for tensile strengths in Table 4.4, although the disparity in strength is somewhat less. Tendon strength is about one order of magnitude less than that of Kevlar or carbon fibers. The specific strength of spider's frame silk is about half that of Kevlar, and it is four times greater than that of high tensile steel. Bone and cuticle are about an order of magnitude less strong than carbon fiber composites. They are, however, somewhat stronger than concrete, but then concrete is not known for its tensile strength. Again, the inference is that biomaterials are relatively low quality designs.

Inspection of the toughness estimates in Table 4.5, however, suggests that this conclusion is not necessarily true. The specific breaking energy of tendon collagen is about one quarter that of Kevlar and about twice that of carbon fiber. The breaking energy of spider's frame silk is six times greater than that of Kevlar and forty times greater than that of carbon fiber. Indeed, silks in general are truly exceptional in their ability to absorb energy (Denny, 1980). Bone and insect cuticle have roughly equal breaking energies to high tensile steel, and their spe-

Table 4.4. *Tensile strength, specific gravity and specific strength data for selected natural and synthetic materials. Data are from Wainwright* et al. *(1982) and Gordon (1978).*

Material	Strength	Sp. gravity	Sp. strength
	(GPa)		*(GPa)*
TENSILE FIBERS			
Tendon (collagen)	0.1	1.3	0.08
Spider's frame silk	1	1.3	0.8
Kevlar	3	1.45	2.0
Carbon fiber	2	2.2	0.9
RIGID MATERIALS			
Bone	0.2	2.2	0.09
Insect cuticle	0.1	1.2	0.08
Carbon fiber composite	1.6	1.8	0.9
High tensile steel	1.5	7.8	0.2
Glass	0.05	2.5	0.02
Concrete	0.004	2.5	0.002

cific breaking energies exceed that of high tensile steel by a factor of about three. The tensile toughness of bone and of cuticle exceed those of glass and concrete by about two and three orders of magnitude respectively. Clearly, in this fracture toughness comparison, the biomaterials no longer appear to be inferior designs, and this suggests that fracture toughness is probably closer to the functional attribute actually selected by evolutionary processes than either stiffness or strength. This should not be taken to mean that stiffness and strength are unimportant, but that the ability to resist fracture may be even more important. Vertebrate tendons and spider's webs, two of the materials discussed above, function in mechanisms that require the absorption of large quantities of mechanical energy, and both materials will benefit if the energy to break is increased. That is, the animals will need less protein to build tendons or webs, and the energy saved will be available for other biological functions. In both cases, this capacity is achieved by combining reasonable stiffness and reasonable strength with a high extension to fracture, making the area under the stress-strain curve (i.e. the energy to break) very large.

Many engineering materials, like carbon fibers, have high stiffness but

Table 4.5. *Toughness, specific gravity and specific toughness data for selected natural and synthetic materials. Toughness is expressed as the energy to break, determined by the area under a tensile stress-strain curve to the point of fracture. Data are derived from infromation in Wainwright et al. (1982) and Gordon (1978)*

Material	Breaking energy $(MJ\,m^{-3})$	Sp. gravity	Sp. breaking energy $(MJ\,m^{-3})$
TENSILE FIBERS			
Tendon (collagen)	6	1.3	5
Spider's frame silk	150	1.3	120
Kevlar	30	1.45	20
Carbon fiber	6	2.2	3
RIGID MATERIALS			
Bone	6	2.2	3
Insect cuticle	4	1.2	3
High tensile steel	6	7.8	1
Glass	0.04	2.5	0.02
Concrete	0.002	2.5	0.0001

at the expense of extensibility, and thus have low toughness. Other materials have high toughness but at the expense of stiffness. Polymeric networks with rubber-like elastic properties, for example, have amongst the highest energies to break of all materials, and yet a vertebrate with rubber bones or tendons would certainly not function well in locomotion. Thus, as expected, the identification of important mechanical properties is complicated. Even if we can identify a mechanical attribute or measure a mechanical property for which the biomaterial is truly exceptional, we can not be certain that this property alone provides the single mechanical criterion which directed the evolution of the biomaterial. Indeed, we may need to identify the several important mechanical properties and will want to develop tests to describe how the balance of these properties affects the quality of the mechanical design. Thus, the real challenge of evaluating the quality of a biomaterial is first to truly understand its function(s), and then to express this understanding in an appropriate set of mechanical properties that can be used to quantify the function.

4.1.5   Structure-property relationships: the essence of design

The focus to this point has been on mechanical properties, but in evaluating design we must consider the relationship between the material structure and mechanical properties. Engineers have studied structure:property relationships extensively and have formulated a set of *Design Principles* that establish appropriate structure-property relationships for well designed materials (Gordon, 1978; Wainwright *et al.*, 1982). Rigid biomaterials, for example, require stiffness, strength, and toughness. Stiffness and strength are easily achieved by having molecules that are linked together by strong covalent or ionic bonds, usually in the form of a regular crystalline lattice. However, such strongly bonded structures are often extremely brittle; that is they lack toughness. There appear to be two basic design strategies for achieving toughness in rigid materials. One involves the use of ductile, metallic materials, but this design strategy is not seen in biology, likely because high temperatures are needed in metal processing (an insurmountable cost of evolution?). The other involves the use of fiber-reinforced composites, in which stiffness and strength are obtained from stiff, usually brittle, reinforcing fibers (e.g. glass or carbon fibers), and toughness is increased by embedding these brittle fibers in a matrix that provides mechanisms to control crack growth. This strategy is used to some extent in virtually all biomaterials.

If we were to look at the microstructure of the rigid biomaterials found in organisms, such as insect cuticle, wood, keratin, and bone, we would see classic, composite structures. Insect cuticle contains stiff fibers formed from the polysaccharide chitin, which are embedded in a protein matrix; woody plant cell walls contain cellulose fibers embedded in a matrix of other complex organic polymers; keratin contains rod-like protein fibers embedded in a cross-linked protein matrix; and bone contains needle-like crystals of hydroxyapatite embedded in a matrix formed by collagen fibers. In other words, these biomaterials have what appear to be appropriate structures, and therefore, they look like good designs. Similar comparisons can be made for tensile fibers, like silk and collagen, and for other classes of biomaterials. They seem to comply with engineering design principles and therefore, like the rigid biomaterials described above, they very likely are good mechanical designs. Although this conclusion is certainly correct, it is somewhat unsatisfying to conclude that biomaterials 'look' like good designs. If we want to really understand the quality of design, particularly with respect to optimization and evolution, we need much more specific information. In particular, we need to know the consequences that changes in material structure will have on functional mechanical properties so that we can discover how properties are optimized. Again, because engineers have been involved in the process of designing composite materials, we can benefit from their experience in attempting to analyze the design of structural materials.

In all cases, the engineering design of materials has involved two stages, (1) the development of analytical models to explain structure-property relationships and (2) the synthesis of model materials in which structure-property relationships can be tested empirically. Composite materials provide excellent examples which have been documented extensively in the engineering literature (Gordon, 1968; Piggott, 1980; Wainwright *et al.*, 1982). For composites made of continuous fibers it is relatively easy to predict stiffness in terms of simple mixture models, in which fiber and matrix stiffnesses and volume fractions are the major variables. Fiber orientation also can be handled relatively easily. The strength and toughness of composites, however, are more difficult to analyze because of the complexity of the mechanisms involved in controlling crack growth in a composite. Variables such as the time-dependent, energy absorbing properties of matrix materials and the strength of the adhesion between matrix and fibers become important. For composites made from discontinuous fibers, a situation common in biomaterials, the transfer of stress from fiber to matrix and vice versa becomes crucially important, and this process is determined by the shape of the fibers, usually expressed as the aspect ratio (length ÷ diameter), as well as the properties of the fiber-matrix interface. Thus, composite materials design is very complex, and invariably an analytical approach combined with extensive empirical testing of individual components and their interaction with each other are essential for refining final designs.

The evaluation of biological materials is complicated by the difficulty of isolating and testing individual fiber and matrix components and of establishing the details of their interactions. In addition, it is generally not possible to construct model materials in which fiber volume fraction, fibre orientation, fibre aspect ratio, fibre-matrix adhesion, etc. can be altered to reveal the mechanical consequences of variation in composite structure. Consequently, in very few instances do we have legitimate assessments of the quality of design in a biomaterial, and in no case is this assessment complete enough that we can claim to really understand the total design of the material. The following examples illustrate systems in which some aspect of design has been evaluated either (1) through the empirical testing of model materials in which structural variables are altered, (2) through empirical testing of natural materials which exhibit a range of structural properties, or (3) through a purely analytical approach.

Koehl (1982) presents one of very few attempts to make model biomaterials in her analysis of spicule-reinforced connective tissues. Many lower animals build modestly stiff structural materials by reinforcing soft connective tissues with rigid, inorganic inclusions, called spicules, made either from calcium carbonate (calcite or aragonite) or amorphous silicon dioxide. Koehl isolated spicules from a variety of animals, and then added them to melted gelatin (Maid Marian brand, raspberry flavour) to make analogues of the natural, spicule-reinforced connective tissues. By correlating the stiffness of the model material with structural parameters such as spicule volume fraction, spicule aspect ratio, spicule length, etc., she was able to provide reasonable explanations of the mechanical properties of the natural, spicule-reinforced tissues that she studied.

Currey (1969) used the natural variation in the mineral content of the bone in rabbit metatarsals to investigate the consequences of varying the volume fraction of the reinforcing fibre on the mechanical optimization of this composite material. The mineral content, as measured by the ash content, varies between about 65 and 70% of the bone weight, with a mean value of about 67%. Currey tested many bone samples and correlated stiffness, strength and a measure of toughness with ash content, as shown in Fig. 4.3. Both stiffness and strength increase linearly with increasing ash content, but interestingly, the toughness reaches a maximum at an ash content of about 67%. Thus, the mean ash content coincides with the maximum in toughness, and it is tempting to speculate that the fracture toughness of bone has been maximized by adjusting the volume fraction of hydroxyapatite crystals in bone. Unfortunately, bone exhibits several levels of composite-like organization, and it is not yet possible to relate these changes in ash content to any specific aspect of bone structure.

Denny (1976) used a simple analytical model to demonstrate the optimization of spider frame silk. The silk fibers in a web are loaded at right angles to their long axes, shown in Fig. 4.2, and in this situation the load (F) supported by the silk fiber depends on both the strength and the extensibility of the silk. The tension (T) which develops in the silk fiber is determined by the magnitude of the load and the deflection angle (θ). For materials with a low extension to failure the deflection angle

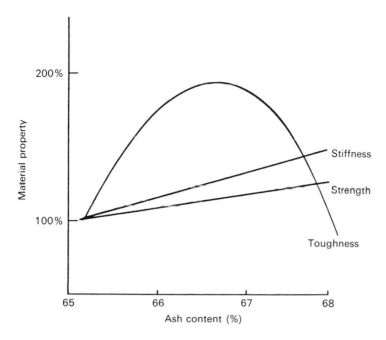

Fig. 4.3. The effect of mineral content, expressed as % ash content, of rabbit metatarsal bone on the relative magnitude of the stiffness, the strength and the toughness. Based on the data of Currey (1969); redrawn from Wainwright *et al.* (1982).

will be small and the tension in the silk strand will be much larger than the applied load. Extensible fibers, however, can rotate to larger angles, coming closer to alignment with the load, and the load and fiber tension become more similar. Thus, more extensible fibers should support larger loads. But as the fiber stretches the cross-sectional area of the fiber decreases, and thus the fiber cross-section available to support the tension decreases. Therefore, there should be an optimum extensibility that allows the greatest load to be supported for a given value of strength. For a fiber that deforms with constant volume this optimum occurs at an extensibility of about 40% (Denny, 1976), and this is very close to the mean extensibility of spider's frame silk. This gives us further evidence that the silk is indeed a very good, if not optimal, design. The structural basis for silk's properties is understood in general terms (Gosline *et al.*, 1986), but a great deal remains to be learned. The silk protein forms micro-crystalline structures that interconnect and reinforce amorphous regions of the protein molecules to form a rigid polymer network. The balance of stiffness, strength, toughness and extensibility necessary for the design of this material is, we believe, due largely to the appropriate arrangement of blocks of crystal-forming and non-crystalline sequences of amino acids in the backbone of silk proteins. Thus, if modern molecular techniques can provide details of the sequence of silk proteins, and if we can correlate differences in sequence structure with mechanical properties, we may obtain a reasonably complete understanding of mechanical design in this class of biomaterials.

4.1.6   Structure-property relationships in soft-biomaterials

Although much research remains to be completed before we can claim to understand the design of any of the stiff or strong biomaterials, at least, on the basis of experience to date, we can use existing engineering theory as a guide for this endeavour. If one looks at the vast variety of animals that exist in nature, however, we find that many, if not most, animals have mechanical support systems based on hydrostatic principles, and hydrostatic skeletons are typically constructed from very low stiffness, and often quite weak biomaterials: the socalled soft-biomaterials or connective tissues. These materials exhibit a wide variety of functional properties and structural motifs that make the analysis of their structure-property relationships extremely complex. Connective tissues are frequently subjected to very large and often multi-axial deformations in their normal use. Stress-strain curves are virtually always non-linear; so the stiffness or modulus of elasticity varies strongly with load and extension. Further, mechanical properties are usually anisotropic and are often strongly time-dependent. Connective tissue structures are equally complex. They contain multi-directional fiber lattices, often with more than one type of fiber (e.g. collagen and elastin fibers in mammalian tissues), embedded in complex, highly hydrated polymeric matrices. In addition fiber orientation and the fiber contour (i.e. crimping or waviness) change dramatically when connective tissues are placed under load. All this makes the quantification of functional or structural attributes extremely difficult; in fact, in many cases appropriate properties have not yet been defined to measure connective tissue attributes.

The elastic tissues of the mammalian arteries provide one of the best understood examples of a complex, connective tissue. In addition to providing a modestly efficient elastic-recoil system, as described previously, arterial elastic tissues must allow for the uniform expansion of the arteries without allowing the formation of local dilations that could

lead to the formation of aneurysms (Gosline, 1980). Uniform expansion
of an elastic tube, like an artery, is achieved by having a non-linear,
J-shaped, stress-strain curve for circumferential expansion, in which the
incremental stiffness increases rapidly with expansion over the physiological range of blood pressures (Bergel, 1961). Thus, the local curvature of
the stress-strain curve is an important mechanical property. However,
the inflation of a cylindrical pressure vessel involves a multi-axial deformation, including strains in longitudinal, circumferential and radial
directions, as well as interactions (i.e. Poisson's ratios) between these
strains. Thus, the development of appropriate properties to quantify
the functional attributes of arterial elastic tissues is extremely complex.
Several attempts have been made to create new mechanical properties
to quantify the shape of stress-strain relations in arterial elastic tissues.
One involves the fitting of strain energy functions (W, strain energy per
unit volume) to polynomial curves (Patel & Vaishnav, 1972), as shown
below.

$$W = Aa^2 + Bab + Cb^2 + Da^3 + Ea^2b + Fab^2 + Gb^3 \qquad (4.1)$$

where a and b represent the strains in the circumferential and longitudinal directions, and A, B, ...G are material constants or mechanical
properties. Although this equation can accurately describe the measured stress-strain relationships of arterial elastic tissues, seven material
properties are required to do this, and there does not seem to be any
simple way to associate this multitude of properties with functionally
important mechanical attributes or with tissue structure. Fung *et al.*
(1979) described a more useful, exponential expression in a paper entitled 'Pseudoelasticity of arteries and the choice of its mathematical expression'. This title clearly expresses the nature of the problem. First,
the mechanical behaviours of arteries and other connective tissues are
so complex that the term elasticity is not deemed appropriate (Fung *et
al.*, 1979; Fung, 1981), and the term pseudoelasticity is suggested as an
alternative. Second, there is choice in the mathematical expressions that
can be invented to quantify pseudoelasticity. The exponential expression
chosen represents the strain energy function (W) as

$$W = C/2\exp[A_1(a^2 - a^{*2}) + A_2(b^2 - b^{*2}) + 2A_4(ab - a^*b^*)] \qquad (4.2)$$

where a and b are circumferential and longitudinal strains, a^* and b^*
are corresponding strains at some reference physiological state, and A_1,
A_2 and A_4 are material constants. Now, with this simpler expression
the material constants, or mechanical properties, can be interpreted in

relatively simple terms. A_1 is a measure of the curvature of the circumferential stress-strain relationship, A_2 is a measure of the curvature of the longitudinal relationship, and A_4 is a measure of the interaction between the two directions. The measured properties depend to some degree on the reference state selected, but at least it is possible to contemplate the possibility of using these mechanical properties to evaluate the structure-property relationships and the quality of mechanical design in elastic arteries. Unfortunately, these analyses remain to be completed. Thus, the analysis of mechanical design in soft-biomaterials remains wide open. In many cases we will not be able to rely on existing engineering theory to provide direction, and it seems likely that biomechanics will have to break new ground to provide a quantitative basis for the science of soft-engineering.

4.2   CONCLUSIONS

The goal at the start of this essay was to discover if biomaterials represented high-quality or optimal mechanical designs. The answer is that, although many biomaterials look like good designs, in most cases we simply do not have adequate information to document the quality or the optimization of biomaterials. For many rigid biomaterials with clearly defined functions it appears that we can rely on existing engineering theory to provide the properties and the analytical methods that will ultimately lead to an understanding of structure-property relationships and the matching of mechanical properties to functional requirements, but much research remains to be completed before the mechanical design of any biomaterial is fully understood.

It is clear, however, that elastic efficiency, or resilience, is not the only, or even a pre-eminent, mechanical property for these analyses. The functional requirements of structural biomaterials are extremely complex, and thus, improving design quality may require the balancing of many properties. For materials that function in energy exchange mechanisms elastic efficiency may provide one of these properties, but it is not necessary that efficiency be maximized. Indeed, some materials may function best when resilience is minimized. The use of efficiency-like effectiveness ratios to gain insights into the processes that direct the evolution of biomaterials is, however, severely limited by our inability to measure complex biological input costs and our inability to assess the effect of changes in mechanical design on animal fitness. At present the best that biomechanics can hope to provide is an evaluation of design quality, but many challenges remain before even this goal can be achieved. The mechanical behaviours and the structures of soft-biomaterials are so much more complex that in many cases it is not possible to use existing engineering theory. We, as biomechanics, will therefore need to develop new mechanical and structural properties that adequately reflect the structure-function relationships of these complex materials before we can even begin to assess the quality of mechanical design in soft-biomaterials. This is a place where biologists, through their investigation of the exceptional diversity of soft-engineered structural systems in organisms, can make substantial contributions to the future directions of engineering.

4.3   REFERENCES

Alexander, R. McN. (1984) Elastic energy stores in running vertebrates, *Amer. Zool.* 24: 85–94.

Alexander, R. McN. (1988) Elastic Mechanisms in Animal Movement, Cambridge Univ. Press, 150 pp.

Bergel, D.H. (1961) Static elastic properties of the arterial wall, *J. Physiol.*, **156**: 445–57.

Currey, J.D. (1969) Mechanical consequences of varying the mineral content of bone, *J. Biomechanics*, **2**: 1–11.

Denny, M. (1976) The Physical properties of spider's silk and their role in the design of orb-webs, *J. exp. Biol.*, **65**: 483–506.

Denny, M. (1980) Silks - their properties and functions, in: *The Mechanical Properties of Biological Materials*, Vincent, J.F.V. & Currey, J.D. (eds.), 34th Symp. of the Soc. for Exptl. Biol., Cambridge Univ. Press., pp. 247–72.

Fung, Y.C. (1981) *Biomechanics: Mechanical Properties of Living Tissues*, Springer-Verlag, New york, Heidelberg, Berlin.

Fung, Y.C., Fronek, K. & Patitucci, P. (1979) Pseudoelasticity of arteries and the choices of its mathematical expression, *Am. J. Physiol.* **237**: H620-H631.

Gordon, J.E. (1968) *The New Science of Strong Materials*, Penguin Books, Ltd., Harmondsworth, England, 269 pp.

Gordon, J.E. (1978) *Structures*, Penguin Books, Ltd., Harmondsworth, England, 395 pp.

Gosline, J.M. (1980) The elastic properties of rubber-like proteins and highly extensible tissues, in: *The Mechanical Properties of Biological Materials*, Vincent, J.F.V. & Currey, J.D. (eds.), 34th Symp. of the Soc. for Exptl. Biol., Cambridge Univ. Press., pp. 331–58.

Gosline, J.M., DeMont, M.E. & Denny, M.W. (1986) The structure and properties of spider silk, *Endeavour*, **10**: 37–43.

Jensen, M. & Weis-Fogh, T. (1962) Biology and physics of locust flight. V. Strength and elasticity of insect cuticle, *Phil. Trans. Roy. Soc. Lond., B* **245**: 137–69.

Ker, R.F. (1981) Dynamic tensile properties of the plantaris tendon of sheep (Ovis aries), *J. exp. Biol.*, **93**: 283–302.

Koehl, M.A.R. (1982) Mechanical design of spicule-reinforced connective tissues: stiffness, *J. exp. Biol.*, **98**: 239–67.

Patel, D.J. & Vaishanv, R.N. (1972) The rheology of large blood vessels, in: *Cardiovascular Fluid Dynanics*, D.H. Bergel, ed., Vol. 2: 2–65.

Piggott, M.R. (1980) Load-Bearing Fibre Composites, Pergamon Press, Oxford, 277 pp.

Vincent, J.F.V. (1982) Structural Biomaterials, John Wiley & Sons, New York, 206 pp.

Vincent, J.F.V. & Currey, J.D. (1980) *The Mechanical Properties of Biological Materials* Cambridge Univ. Press, 513 pp.

Wainwright, S.A (1988) *Axis and Circumference*, Harvard Univ. Press, 132 pp.

Wainwright, S.A., Biggs, W.D., Currey, J.D. & Gosline, J.M. (1982) *Mechanical Design in Organisms*, Princeton Univ. Press, Princeton, 423 pp.

Weis-Fogh, T. (1972) Energetics of hovering flight in hummingbirds and Drosophila, *J. exp. Biol.*, **56**: 79, 104.

5

Efficiency and optimization in the design of skeletal support systems

A. A. BIEWENER
and
J. E. A. BERTRAM

5.1 ABSTRACT

Optimization approaches, if properly applied and interpreted, can be of value in identifying and evaluating critical features of the design of skeletal support systems. However, the use of 'efficiency' (defined as strength per unit mass) to characterize the quality of skeletal design is often misleading or unwarranted because of (i) its lack of correspondence to thermodynamic efficiency and (ii) mechanical properties other than strength (e.g. toughness, stiffness, strain energy recovery) are often important in the design of skeletons. Instead, less ambiguous terms, such as specific strength or specific stiffness, should be used. In this Chapter, we review cases in which design to maximize specific strength provides a good explanation of certain features of bone shape and organization in the limb. However, when the biological roles of bone elements differ significantly, evaluation of the material and/or structural properties of bone by any one criterion, such as failure strength, is insufficient. In this case, optimization of competing functional properties must be considered. At a structural level, optimization of bone shape for strength versus loading predictability appears likely in the long bones of mammals. We also show that a trade off exists in the design of tendon for effective elastic energy recovery in large kangaroos versus acceleration and strength in mush smaller kangaroo rats. More generally, thick tendons appear to reflect the importance of stiffness for motor control of muscle length change and motion at a joint. Though optimality approaches help to evaluate the interaction of such constraints on design, it should be recognized that such approaches largely ignore historical constraints on form and function.

5.2   INTRODUCTION

In this Chapter we address the concepts of efficiency and optimization in the design of skeletal support systems, particularly in regard to their use for identifying those factors or properties which underlie the observed design and are critical to its function. In other words, what properties should a 'good' skeleton possess? What is a good design? Before these questions can be answered, however, the functional requirements of the skeletal system must be identified. A difficulty in establishing the functional role of any structure is the possibility that a structure may have multiple functions. For instance, although a clear functional requirement of any skeleton is not to break (in which strength is the relevant material property), other properties such as stiffness or the ability to absorb energy upon impact are also likely to be important to its design and function.

In contrast to the definition of thermodynamic efficiency, skeletal efficiency traditionally has been defined as the strength of a support element per unit mass, which in effect represents a benefit:cost ratio the ability to support load versus the cost to build and maintain the element. This ratio, in fact defines the specific strength of a structure. Therefore, because 'skeletal efficiency' does not conform to the definition of thermodynamic efficiency and mechanical properties other than strength (e.g. toughness, stiffness, resiliency) may be important in the biomechanics of skeletal support, specific strength is preferred instead. The strength of a structure is defined as the maximum stress (force/cross-sectional area) that a structure can support before failing, and represents the most common criterion used to evaluate skeletal design (Alexander, 1981, 1983; Biewener, 1989a,b; Currey, 1984), with the assumption that changes in material organization or shape that increase a structure's specific strength will be favored. A notable exception is McMahon's (1975) elastic similarity model, which attempts to explain the scaling of skeletal form primarily on the basis of maintaining similar elastic deformations.

Optimization, on the other hand, defines a process by which two or more competing functional requirements of a system are maximized. Optimality approaches recognize that structures may perform multiple functions and are meant to test the relative importance of differing, and

Table 5.1. *Effect of percentage mineralization on the material properties of bone (from Currey, 1979)*

Skeletal element	Mineralization (%)	Work of fracture (Jm^{-3})	Strength (MPa)	Stiffness (GPa)
Deer antler	59	6186	179	7.4
Cow femur	67	1710	247	13.5
Whale bulla	86	200	33	31.3

potentially competing, functions. Intrinsic to the application and interpretation of optimality approaches are two key assumptions: (1) that all relevant functions are identified, and (2) that these functions can be quantified (as well as the optimization function itself). These assumptions must be recognized as limiting the predictive value of any optimization model. In addition, it should be noted that optimality approaches traditionally ignore historical constraints on the form and function of a system, assuming adaptive equilibrium between the organism and its environment (Lauder, 1981; Liem and Wake, 1985).

In this Chapter, we show that although specific strength explains well several features of the skeleton (shape and reduced element number), other functional requirements may represent important trade-offs in the design of these elements versus their strength. Two competing functional requirements identified are: (1) increased loading predictability in mammalian long bones, and (2) increased elastic strain energy recovery in tendons. Though beginning our discussion with the material properties of bone, we intend to focus in this Chapter on the organization and function of skeletal tissues as structural elements (see Gosline for a discussion of the mechanical properties of biological materials, Chapter 4).

5.2.1 Material properties of bone: optimization of strength, stiffness and toughness?

At a material level, most bones possess quite uniform mechanical properties (Currey, 1984). Over a broad range of body size the strength of bone (stress at failure) does not change significantly (Biewener, 1982). Correspondingly, mineralization levels rarely vary beyond the range of 63 to 68%. Above this level of mineralization, the toughness (energy-absorbing capacity) of bone declines with little increase in strength; whereas with decreased mineralization, the stiffness of bone is lowered (Currey, 1981). This apparent uniformity of bone tissue properties in differing skeletal elements of diverse vertebrate taxa suggests that there has been strong selection favoring a narrow mineralization range for bone, optimizing a bone's ability to be stiff and strong, as well as tough.

When functional requirements vary from that of body support, however, the material properties of a bone may differ considerably. The change in material properties again is largely due to a change in the relative amounts of mineral and organic matrix (principally collagen) present in the bone. Currey (1979; Table 5.1) showed clearly how increased mineralization of the auditory bulla of the whale maximizes stiffness, providing high acoustic impedance for effective sound transmission. The increase in stiffness is at the expense of low strength and toughness, properties unlikely to be of functional consequence in the skull of an aquatic animal. In contrast, bone from a deer antler has a high energy-absorbing capacity associated with a much lower mineralization level, presumably of functionally importance for withstanding impact loading. The femur possesses an expected intermediate level of mineralization as a limb support element, reflecting its need to be strong and stiff, as well as tough. The greatly differing mineralization levels and mechanical properties of these three skeletal elements show that considerable change in bone mineralization is possible, if selectively advantageous.

5.2.2 *Bone shape and bone number in relation to the optimization of specific strength*

Optimization of bone shape and the reduction of the number of limb support elements for maximizing specific strength and lowering limb inertia are well recognized design features of the vertebrate skeleton. With an increase in the length of a limb bone (most often associated with increased body size within a lineage), the problem of support against bending loads is greatly exacerbated. Indeed, bending accounts for 70 to 90% of the stress developed during locomotion in most long bones of both large and small animals (Bertram and Biewener, 1988). Because of the uneven distribution of stresses in a beam loaded in bending, in which stresses are greatest at the outer surfaces of the beam and least at its center (zero along the neutral axis), a tubular cross-sectional shape provides the best form to resist bending. By distributing bone tissue away

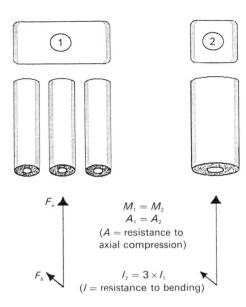

Fig. 5.1. Schematic drawing of limb support elements (bones) showing how specific strength (strength/mass) increases with a reduction in the number of support elements at a given level within a limb. F_b acts transversely in the anteroposterior plane (M: bone mass; A: bone area; I: second moment of area in the anteroposterior direction). This pattern is observed in ungulate phylogeny.

from the neutral plane of bending, the resistance of bone to bending (second moment of area) is increased per unit mass of bone tissue (Currey, 1984). It should be noted again that specific features of a bone's cross-sectional shape (extent of cortical thinning relative to inner and outer diameters of the bone) may be important with regard to trade-offs among differing structural properties of the bone and modes of failure (Currey and Alexander, 1985).

The decrease in the number of distal limb support elements associated with the evolutionary increase in size within ungulates (and to a lesser extent, other groups as well) similarly reflects a design to increase the specific strength of the skeleton (Fig. 5.1). Here (as with shape), the advantage to the animal is not only the economy of tissue required for support, but arguably the energy saved by reducing the inertia of swinging its limbs during locomotion. With the reduction of skeletal elements, muscle mass is also reduced. The functional trade-off, however, is reduced mobility of these elements and a more restricted range of locomotor performance.

5.2.3   Stress similarity in the design of the skeleton

The generally uniform mechanical properties of vertebrate bone and the force-generating capacity of striated skeletal muscle (reflecting its conservative organization of actin and myosin: Close, 1972; Biewener *et al.*, 1988), have led us to propose the simple hypothesis of 'stress similarity' in the design of muscle and bone in relation to the mechanics of support. That is, we propose that similar peak stresses (specific to each kind of support element) act under equivalent locomotor conditions in different species (Biewener and Taylor, 1986; Biewener *et al.*, 1988; Perry *et al.*, 1988; Rubin and Lanyon, 1984; Taylor, 1985). Accordingly, this hypothesis argues that the amount, form and organization of the skeleton will be adjusted in relation to the overall functional capacity of the animal, such that a similar factor of safety is achieved for each type of support element (muscle, tendon, and bone). It should be noted that, as with the previous considerations of bone shape and number, this hypothesis assumes specific strength, or peak stress, to be the limiting design criterion of the skeleton.

In support of stress similarity, peak stresses acting in the long bones of mammalian species during strenuous activity (fast galloping and jumping) appear similar over a wide range of size, maintaining a safety factor of between two and four relative to the compressive failure stress of bone (Fig. 5.2). Further, in two species of equal size but differing locomotor ability (goat and dog), we found that similar stresses acted in the radius and tibia of each species when compared at equivalent points of their gait (trot-gallop transition and top galloping speeds: Biewener and Taylor, 1986; Rubin and Lanyon, 1982). Similar peak stresses were achieved by increased bone area and second moment of area in the dog compared to the goat, enabling the dog to attain an absolute speed 1.7 times faster than the goat. Finally, we have shown that similar stresses (70 kPa; about 33% of peak isometric stress) act in the ankle extensor muscles of the quadrupedal white rat (galloping) and bipedal kangaroo rat (hopping) when each moves at its preferred speed (1.5 m s^{-1} for both species, Perry *et al.*, 1988).

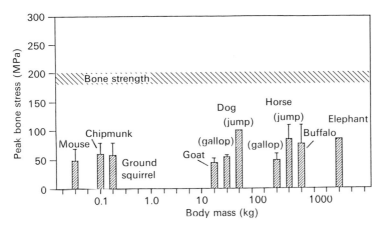

Fig. 5.2. A histogram of peak stresses acting in limb bones of different sized mammalian species during stenuous activity (galloping and jumping). The fairly uniform peak stresses observed in these different sized species indicate that the mammalian skeleton maintains a safety factor of about three relative to the compressive failure strength of bone (hatched bar). Error bars reflect the SD in peak stress for different elements within each species. Data are for: mouse (Biewener, unpublished); ground squirrel and chipmunk (Biewener, 1983); goat (Biewener and Taylor, 1986); dog (Rubin and Lanyon, 1982); kangaroo (Alexander and Vernon, 1975); horse (Biewener *et al.*, 1983, 1988); buffalo and elephant (Alexander *et al.*, 1979).

5.2.4 Design for predictability versus strength

Up to this point, we have discussed features of skeletal design that are well explained by the traditional view of selection favoring an increase in specific strength and the maintenance of stress similarity. Many long bones of terrestrial mammals however are curved along their length (Fig. 5.3). In some bones, this curvature can be quite large. Though longitudinal curvature was originally thought to neutralize, or counteract, externally applied bending moments (Frost, 1964; Pauwels, 1980), in several long bones studied the bone's longitudinal curvature promotes, rather than diminishes, bending-induced stresses developed within the bone (Biewener *et al.*, 1983, 1988; Biewener and Taylor, 1986; Lanyon and Bourn, 1979; Rubin and Lanyon, 1982). Evaluated in terms of (maximizing) specific strength, bone curvature thus presents a paradox in design, as considerably more bone tissue is required for a given level of stress (Fig. 5.4a).

We have recently argued that this curvature may function to increase the loading predictability of long bones (Bertram and Biewener, 1988).

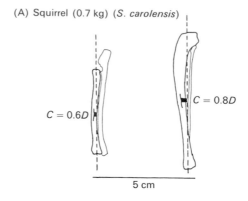

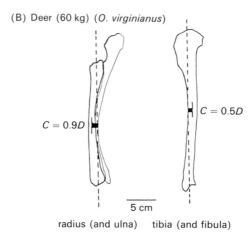

radius (and ulna)   tibia (and fibula)

Fig. 5.3. In medial view: the radius and tibia of (A) a squirrel and (B) a deer, showing the antero-posterior curvature of these bones (the ulna and fibula are shown by thinner lines). The curvature (C) of each bone is indicated (dark stippled bar), measured as the moment arm of axial force normalized to the diameter (D) of each bone at its midshaft.

Improved predictability of loading could be of value for support elements exposed to variable dynamic loads. We developed a model to determine how the strength, or 'load carrying capacity' of the bone, might inter-act with predictability of loading to begin to test this idea. Curvature (C) provides an intrinsic preferred (controlled) plane of bending for the element, in the direction of its curvature. As expected, predictability of loading direction increases, whereas load carrying capacity decreases,

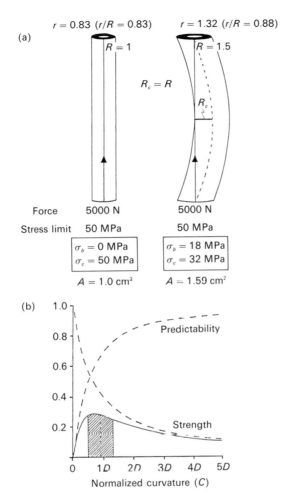

Fig. 5.4. (a) Effect of curvature on bone stress (R: external radius; r: internal radius; R_c: moment arm due to bone curvature; A: bone cross-sectional area; σ_b: stress due to bending; σ_a: stress due to axial compression). Force and stress limits selected are arbitrary. (b) A graph showing the increase in loading predictability versus a decrease in load-carrying capacity (strength) with increased bone curvature normalized to bone diameter (D) (based on a model from Bertram and Biewener, 1988). The product interaction of these two factors is maximal at a curvature equal to 0.6 times the bone's diameter. The hatched region reflects the range of curvatures (mean: 0.85) exhibited by the radius in mammals ranging from 0.01 to 50 kg in body mass.

with increased curvature (Fig. 5.4b). The interaction between load carrying capacity and loading predictability was optimized at a curvature equal to 0.6 times the diameter (D) of the support element in our model. This value, in fact, matches closely the observed curvature of the radius (forearm bone) for a diverse range of mammals (Bertram, 1988). Loading of this bone during locomotion corresponds best to the conditions of our model. Interestingly, in the largest mammalian species (> 50 kg) curvature of the radius diminishes considerably, becoming quite straight in the largest species, as would be expected if the ability to maintain strength is the critical factor at an extreme in body size. In other experiments, in vivo bone-strain data recorded during locomotion in fore limb and hind limb bones of the horse indicate that loading predictability is greater in the curved radius and tibia, compared to the much straighter metacarpus and metatarsus (Biewener *et al.*, 1983, 1988), especially during acceleration and deceleration, as in the take-off and landing of jumps. The greater variation in loading of these straight bones corresponds to a higher incidence of fracture of these elements in racehorses (Currey, 1981), indicating the potential importance of control of loading direction when an animal nears the limit of its performance range. A third piece of evidence supporting the functional value of curvature to restrict variability of stress distribution within a bone is the observation that a bone loses, or fails to develop, its normal curvature if the limb is deprived of weight support and locomotor function during growth (Biewener and Bertram, unpublished data; Lanyon, 1980).

Evaluated solely in terms of maximizing specific strength of the bone, curvature can not be shown to be advantageous. However, by restricting the range of bending orientation to which a bone is subjected, the number and mass of muscles acting about the bone may also be reduced. Considered together, the savings in muscle mass (and the associated energy cost of operating the muscles) may outweigh the increase in bone mass. Even so, it seems likely that improved predictability of loading (and hence, maintained stress distribution within the bone) represents an important design criterion in the context of skeletal remodeling in response to mechanical forces (Lanyon, 1984; Lanyon and Rubin, 1985), which should be considered along with that of specific strength. Though simple, our model shows how these two competing design criteria can be optimized to achieve a form that is generally characteristic of long bone elements in many different mammals.

5.2.5   Design for elastic energy savings versus strength

Not only may control for predictability of loading (bending) direction represent a trade-off in design for strength, but effective elastic strain energy recovery in tendons (and ligaments) also favors a design to reduce the safety factor of the tendon. Elastic energy recovery depends on large strains acting in tendons during the activity in which energy saving is believed important to the animal (typically locomotion: e.g. foraging and/or migration). This results from the squared relationship between strain energy per unit volume (U) and strain (ε), where $U = \frac{1}{2}E.\varepsilon^2$ for linearly elastic materials and E is Young's modulus. Mammalian tendons, in fact, behave as linear elastic elements over much of their strain range and have resiliencies as high as 94% (Bennett *et al.*, 1986). For tendons of equal length and transmitted force, a tendon that is only 50% as thick can generally store four times more energy. On a mass-specific basis, the increase in energy recovery is 16–fold for the thinner tendon! Hence, long thin tendons are the best form for storing and recovering strain energy.

Competing with the design of a thin tendon for high strain energy savings are the requirements for acceleration (strength) and need for effective motor control (stiffness) of the limb. Each of these latter two requirements favors a thicker, and hence, stiffer tendon. Increased tendon thickness means that greater force can be transmitted safely to accelerate the animal. Increased stiffness allows for more precise control of length changes by the muscle at a joint, particularly when the muscle's fibers are short relative to the length of the tendon (Ker, Alexander and Bennett, 1988). Further, because the range of force (or stress) that a tendon must bear is dictated entirely by the maximum force that its muscle can exert, a close match between the design of the tendon to its muscle seems likely.

We have examined the question of design for elastic energy savings versus acceleration ability by comparing a small hopping heteromyid rodent, the kangaroo rat (*D. spectabilis*, 0.1 kg) with the much larger red kangaroo (*M. rufus*, 40 kg). Though at first glance the hind limb morphology and bipedal gait of these two species appear quite similar (despite their difference in size), we find that the structural design of the tendons, muscles and bones in each of these species is strikingly different. When considered in the context of steady speed hopping, the kangaroo rat appears greatly 'overbuilt'. Stresses in its tendons and bones during hopping indicate a safety factor of between eight and ten (Fig. 5.5a), correlated with its inability to recover much strain energy during hopping ($< 14\%$ of the work done per stride: Biewener *et al.*, 1981). Peak stresses in the ankle extensor muscles similarly are quite low during hopping, being 35% of the peak isometric stress of the muscle. However, kangaroo rats demonstrate a considerable ability to jump, attaining heights of at least 50 cm (10 times their hip height). During these high jumps, peak stresses rise to levels consistent with those calculated for other mammalian species under conditions of strenuous activity (safety factor of between two and three; Biewener and Blickhan, 1988). This exceptional ability to jump and sustain the large forces imparted to its limbs (up to 10 times body weight per limb), in fact, has been shown to be critical to the locomotor strategy of the kangaroo rat for effective escape from predation under natural conditions (Webster and Webster, 1980).

In contrast to the smaller kangaroo rat, the red kangaroo possesses much thinner tendons and muscles of smaller fiber cross-sectional area relative to its size (Fig. 5.5). The smaller dimensions of its ankle extensor muscles and tendons indicate that, for equivalent accelerations of its body, the stresses developed in these two elements are likely to be considerably greater. When compared at their preferred hopping speed (kangaroo rat: 1.5 ms^{-1} and kangaroo: 3.9 ms^{-1}), we find this to be the case. Whereas peak stresses in the muscles of the kangaroo rat are only 70 kPa, peak stresses of 180 kPa act in the red kangaroo (R. Kram, personal communication; Fig. 5.5a), close to the peak isometric stress exerted by skeletal muscle (180–240 kPa; Biewener *et al.*, 1988; Close, 1972). Similarly, peak stresses acting in the tendons of the kangaroo hopping at its preferred speed are 55 MPa (safety factor of two), compared to only 8 MPa in the kangaroo rat (Fig. 5.5b).

This 2.5–fold difference in peak muscle stress and seven–fold difference in tendon stress has two important implications. First, the high stresses observed in the kangaroo are consistent with its remarkable ability to conserve energy by elastic strain energy recovery (dependent on high tendon strain, in long tendons), particularly at higher hopping speeds (Alexander and Vernon, 1975; Dawson and Taylor, 1973). Second, the much thinner tendons and smaller muscles of the kangaroo (relative to its size) pose an important functional trade-off for the animal, severely limiting the accelerative capacity of its locomotor system. For a factorial increase in acceleration (steady speed hopping versus maximal jumping) equivalent to that observed in the smaller kangaroo rat, peak stresses

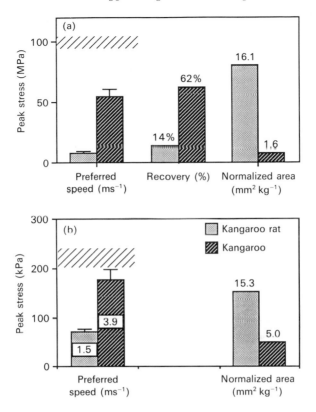

Fig. 5.5. (a) A histogram of peak stress, percentage energy recovery, and normalized area for the ankle extensor tendons of the kangaroo rat vs. the red kangaroo. Peak stresses were those developed at the preferred hopping speed (shown) for each species. The hatched bar indicates the failure strength of tendon. (b) A histogram of peak stress (at preferred speed) and relative fiber cross-sectional area for the ankle extensor muscles of the kangaroo rat vs. the kangaroo. The hatched bar indicates the peak isometric stress exerted by skeletal muscle. (Data for the kangaroo rat from Biewener *et al.*, 1988, and for the kangaroo courtesy of R. Kram, unpublished.)

would exceed the physiological limit of the kangaroo's muscles by more than two–fold and their tendons would rupture if such high forces were exerted (Fig. 5.6). It seems clear that such distinct locomotor capacities suggest differing, competing selective advantages of bipedal hopping to each species in an ecological context. Whereas its low frequency auditory sensitivity (Webster and Webster, 1980) and jumping ability have enabled the kangaroo rat to exploit open territory, elastic energy savings allows the red kangaroo to forage and migrate long distances, seemingly with minimal risk of predation.

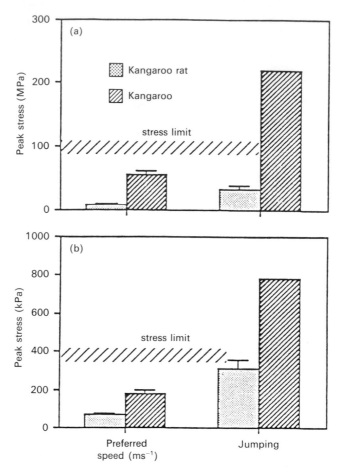

Fig. 5.6. A histogram comparing the performance capacity of the kangaroo rat vs. the kangaroo in terms of (a) peak tendon stress and (b) peak muscle stress. The stresses shown for the kangaroo during jumping were calculated based on the factorial increase achieved by the kangaroo rat (from hopping to jumping) and clearly show the kangaroo's inability to achieve or support such high stress levels, which greatly exceed the stress limit of both its muscles and tendons.

Whether the difference in skeletal design between these two bipedal species reflects mainly the scaling effect of body size remains an open question. Additional studies of other species, particularly large quadrupeds, is needed to test this possibility. Recent data, in fact, suggest that the tendons of animals are generally thicker than would be expected on the basis of strength, let alone effective strain energy recovery (Ker *et al.*, 1988). The basis of this appears to be reduced compliance (increased stiffness) for more effective control of muscle-tendon length changes and joint movement. The small cross-sectional area of the kangaroo's tendons and high strains developed in them during hopping, therefore suggest that elastic energy savings may be favored at the expense of reduced motor control as well.

5.3 CONCLUSION

In this Chapter, we have shown that whereas design to maximize specific strength is clearly important in the skeletal support system of mammals, other properties, such as stiffness, elastic energy recovery and loading predictability, may also be important design criteria as well. In most cases, these represent competing trade-offs in the design of a support element. An optimization approach may be useful in determining the relative importance of each to the design of the element. In addition to evaluating the quality of design, such an approach can also help to identify other factors of functional importance. The use of 'efficiency' to evaluate skeletal design, commonly defined in benefit/cost terms as strength/mass (or specific strength), has no thermodynamic meaning and may misrepresent the functional requirements of the skeleton. Specific strength is commonly the basis for understanding how skeletal dimensions scale with size, why bones possess a tubular shape, and phyletic reduction in support element number. However, stiffness may be equally important to the function of the skeleton, not only as a material property, but as a determinant of a bone's cross-sectional shape. Improved predictability of loading direction and hence, maintained stress distribution within a bone, represent potential advantages to explain the presence of longitudinal curvature in limb bones. As such, we believe that loading predictability represents a second important design criterion versus strength. Finally, effective storage and recovery of elastic strain energy requires that high strains be developed in tendons (or ligaments). Hence, elastic energy recovery represents a third competing factor versus strength. The functional consequences associated with this trade-off in design are dramatically illustrated by the differing locomotor abilities of the kangaroo rat and the much larger red kangaroo. Though both species move by hopping and appear similarly built, the kangaroo rat is designed for acceleration to escape predation, whereas the kangaroo achieves exceptional elastic energy savings, and hence endurance, during hopping but has limited accelerative ability.

Several design trade-offs, therefore, appear to exist within the skeletal support system of mammals, reflecting the differing, and often competing, functions that a skeleton must fulfill. Identifying the functional requirements of a skeleton and quantifying their relative importance as interacting influences on its design can provide valuable insight into the evolution of form and function.

5.4 ACKNOWLEDGEMENTS

We thank Mr Roger Kram for the use of his data for the red kangaroo. Portions of the authors' work discussed in this chapter were supported by NSF grant DCB 85–14899 and a grant from the Whitaker Foundation.

5.5 REFERENCES

Alexander, R. McN. (1981). Factors of safety in the structure of animals. *Scientific Progress* **67**:109–30.

Alexander, R. McN. (1983). *Animal Mechanics*, 2nd edn, Blackwell Scientific. London.

Alexander, R. McN., Maloiy, G.M.O., Hunter, B., Jayes, A.S. & Nturibi, J. (1979). Mechanical stresses in fast locomotion of buffalo (*Syncerus caffer*) and elephant (*Loxodonta africana*). *J. Zool., Lond.* **189**:135–144.

Alexander, R. McN. & Vernon, A. (1975). The mechanics of hopping by kangaroos (Macropodidae). *J. Zool., Lond.* **177**:265–303.

Bennett, M.B., Ker, R.F., Dimery, N.J. & Alexander, R. McN. (1986). Mechanical properties of various mammalian tendons. *J. Zool., Lond.* **209**: 537–48.

Bertram, J. E. A. (1988). The Biomechanics of Bending and its Implications for Terrestrial Support. Ph. D. thesis, The University of Chicago, Chicago, IL.

Bertram, J.E.A. & Biewener, A.A. (1988). Bone curvature: sacrificing strength for load predictability? *J. theor. Biol.* **131**:75–92.

Biewener, A. A. (1982). Bone strength in small mammals and bipedal birds: do safety factors change with body size? *J. exp. Biol.* **98**: 289–301.

Biewener, A.A. (1983). Locomotory stresses in the limb bones of two small mammals: the ground squirrel and chipmunk. *J. exp. Biol.* **103**: 135–154.

Biewener, A. A. (1989a). Scaling body support in mammals: limb posture and muscle mechanics. *Science* **245**:45–8.

Biewener, A.A. (1989b). Mammalian terrestrial locomotion and size. *Bioscience* **39**:776-783.

Biewener, A.A., Alexander, R.McN. & Heglund, N.C. (1981). Elastic energy storage in the hopping of kangaroo rats (*Dipodomys spectabilis*). *J. Zool., Lond.* **195**:369–83.

Biewener, A. A., Blickhan, R., Perry, A. K., Heglund, N. C. & Taylor, C. R. (1988). Muscle forces during locomotion in kangaroo rats: force platform and tendon buckle measurements compared. *J. exp. Biol.* **137**: 191–205.

Biewener, A.A. & Blickhan, R. (1988). Kangaroo rat locomotion: design for elastic energy storage or acceleration? *J. exp. Biol.* **140**:243–55.

Biewener, A. A. & Taylor, C. R. (1986). Bone strain: a determinant of gait and speed? *J. exp. Biol.* **123**: 383–400.

Biewener, A. A., Thomason, J. & Lanyon, L. E. (1983). Mechanics of locomotion and jumping in the forelimb of the horse (*Equus*); in vivo stress in the radius and metacarpus. *J. Zool., Lond.* **201**: 67–82.

Biewener, A. A., Thomason, J. & Lanyon, L. E. (1988). Mechanics of locomotion and jumping in the horse (*Equus*): in vivo stress in the tibia and metatarsus. *J. Zool., Lond.* **214**: 547–565.

Close, R. I. (1972). Dynamic properties of mammalian skeletal muscles. *Physiol. Rev.* **52**:129–97.

Currey, J. D. (1979). Mechanical properties of bone with greatly differing functions. *J. Biomech.* **12**:313–19.

Currey, J. D. (1981). What is bone for? Property-function relationships in bone. In *Mechanical Properties of Bone*, S.C. Cowin, ed. ASME.

Currey, J. D. (1984). *The Mechanical Adaptations of Bone*. Princeton University Press, Princeton, NJ.

Currey, J. D. & Alexander, R. McN. (1985). The thickness of the walls of tubular bones. *J. Zool., Lond.* **206**:453–68.

Dawson, R.J. & Taylor, C.R. (1973). Energy cost of locomotion by kangaroos. *Nature, Lond.* **246**: 313–14.

Frost, H.M. (1964). *The Laws of Bone Structure*. Charles C. Thomas, Springfield, IL.

Ker, R.F., Alexander, R. McN. & Bennett, M.B. (1988). Why are mammalian tendons so thick? *J. Zool., Lond.* **216**:309–24.

Lanyon, L.E. (1980). The influence of function on the development of bone curvature. An experimental study on the rat tibia. *J. Zool., Lond.* **192**: 457–66.

Lanyon, L.E. (1984). Functional strain as a determinant of bone remodeling. *Calc. Tiss. Int.* **36**: S56-S61.

Lanyon, L.E. & Bourn, S. (1979). The influence of mechanical function on the development and remodelling of the tibia: an experimental study in sheep. *J. Bone Jt. Surg.* **61 A**: 263–73.

Lanyon, L.E. & Rubin, C.T. (1985). Functional adaptation in skeletal structures. Ch. 1 in *Functional Vertebrate Morphology* (M. Hildebrand, D.M. Bramble, K.F. Liem & D.B. Wake, eds.), Harvard University Press, Cambridge, MA.

Lauder, G.V. (1981). Form and function: structural analysis in evolutionary morphology. *Paleobiol.* **7**:430–42.

Liem, K.F. & Wake, D.B. (1985). Morphology: current approaches and concepts. Ch. 18 in *Functional Vertebrate Morphology* (M. Hildebrand, D.M. Bramble, K.F. Liem & D.B. Wake, eds.), Harvard University Press, Cambridge, MA.

McMahon, T. A. (1975). Size and shape in biology. *Science* **179**: 1201–4.

Pauwels, F. (1980). *Biomechanics of the Locomotor Apparatus.* Springer-Verlag, Berlin.

Perry, A. K., Blickhan, R., Biewener, A. A., Heglund, N. C. & Taylor, C. R. (1988). Preferred speeds in terrestrial vertebrates: are they equivalent? *J. exp. Biol.* **137**: 207–20.

Rubin, C. T. & Lanyon, L. E. (1982). Limb mechanics as a function of speed and gait: a study of functional strains in the radius and tibia of horse and dog. *J. exp. Biol.* **101**: 187–211.

Rubin, C. T. & Lanyon, L.E. (1984). Dynamic strain similarity in vertebrates: an alternative to allometric limb bone scaling. *J. theor. Biol.* **107**:321–7.

Taylor, C. R. (1985). Force development during sustained locomotion. *J. exp. Biol.* **115**: 253–62.

Webster, D.B. & Webster, M. (1980). Morphological adaptations of the ear in the rodent family Heteromyidae. *Am. Zool.* **20**:247–54.

6

Efficiency in aquatic locomotion: limitations from single cells to animals

T. L. DANIEL

6.1 INTRODUCTION

Efficiency, commonly defined as the rate of useful energy expenditure divided by the total rate of energy consumption, has been used as a measure of performance for many biological systems, ranging from efficiency of conversion of chemical to mechanical energy, or the conversion of one form of mechanical energy to kinetic or potential energy in moving organisms. In animal locomotion, efficiency has been used as a measure of performance with the idea that maximization of this parameter correlates, in some manner, with increased fitness. Thus animal morphologies, limb kinematics, and even tissue elasticity are commonly examined in light of how these phenotypic features may affect the overall efficiency of moving animals (see for example, Alexander and Bennet-Clark, 1977; Cavagna *et al.*, 1977; Lighthill,1975; Wu, 1971; Weihs and Webb, 1983; Blake, 1983). The extent to which natural selection acts on efficiency in animal locomotion remains unclear. Therefore, rather than examining the maximum attainable efficiencies for moving animals, this Chapter is concerned with the limits to efficiency that may be set by both the physics and physiology of movement in animals. I use aquatic locomotion to illustrate such limitations, examining a broad size range of animals which use diverse modes of propulsion. I seek general trends in the kinematic and morphological factors that affect levels of efficiency. I then delve more deeply into the efficiency limits imposed by the muscles and tissues involved in producing movement.

A simple flow diagram of the steps of energy conversion in a moving animal shows the various levels at which we can examine concepts of efficiency in aquatic locomotion (Fig. 6.1). In a general sense, the overall efficiency of the animal is simply the ratio of thrust work done (mean thrust times speed) in moving the animal to the total rate of oxygen consumption. That overall efficiency, in turn, is determined by three steps: (1) the efficiency with which muscles use oxygen in hydrolyzing ATP to generate work; (2) the fraction of muscle work that moves fluid; and (3) the fraction of work done in moving fluid that goes to useful thrust. This third step is often termed the 'Froude efficiency' in aquatic locomotion (Lighthill, 1975; Blake, 1981a,b, 1983). In step (2), some of the contractile energy of propulsive muscles goes to deforming parts of the body involved in thrust production. Some of this work is potentially recoverable as elastic energy recoil (Alexander and Bennet-Clark, 1977) whereas the rest is lost in accelerating the mass of the body part and overcoming viscous damping in the tissues. There is an ample supply of theory and experiment to examine the factors that determine Froude efficiency in aquatic animals and I will review that briefly here. While the implications of energy recoil have received considerable attention in terrestrial systems (Biewener *et al.*, 1981, 1988; Cavagna and Taylor, 1977) far less is known about the determinants of step (2) for aquatic systems. I delve into this issue with a case study shrimp locomotion where more data exist.

6.2   DIVISIONS OF AQUATIC LOCOMOTION

Research on aquatic locomotion can be usefully examined from several points of view. In one case, we might be interested in locomotor performance of animals that move in a relatively steady manner, rarely accelerating, decelerating or turning. Such motions are logically called steady state swimming (Daniel and Webb, 1987) and, for such cases, efficiency is simply defined as above by the product of thrust times swimming speed divided by the total energy input to the organism. Alternatively, there are other situations such as predator-prey interactions in which turns or accelerations dominate the motion. For such unsteady modes of locomotion, such a definition of efficiency may not be a biologically relevant measure of performance. Instead, some other measure, such as the total work done in accelerating the animal divided by the total available energy, may be more meaningful. The total work is simply the product of the mass times the mean acceleration and distance reached during an escape or attack.

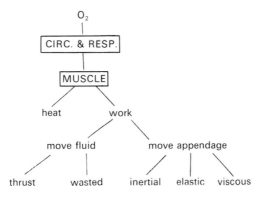

Fig. 6.1. A flow diagram of the steps of energy conversion during locomotion. Oxygen, brought to muscle by circulatory and respiratory systems is converted to contractile work. The efficiency of this conversion is often taken to be 20%. Of the contractile work available, some goes to moving fluids and some goes to deforming tissues. The former yields a fraction of work that goes to useful thrust, called the Froude efficiency. The latter may be stored elastically in the tissues, may be viscously dissipated in the tissues or may go to accelerating the limb.

The various modes of propulsion also serve to delineate studies of aquatic locomotion. Rowing appendages in aquatic insects, fish and small crustaceans, body undulations in sperm and eels, and wing-like devices in penguins, sea turtles and tunas all use subtly different mechanisms in generating thrust and have different determinants of efficiency. While there is a vast literature dealing with such divisions (for reviews see Lighthill, 1975; Wu *et al.*, 1975; Webb and Weihs, 1983; Blake, 1983), I will only examine just a few examples of these steady state swimming modes both to review the basic literature on aquatic propulsion and to set the stage for examining how the mechanics of propulsive muscle may limit overall efficiency in locomotion. The review is meant to be neither exhaustive nor all inclusive.

6.3   EFFICIENCY IN STEADY STATE SWIMMING

6.3.1   Locomotor efficiency in undulators

Undulatory locomotion occurs over an impressive size range of animals, ranging from the sinuous movements of sperm flagellae only 40

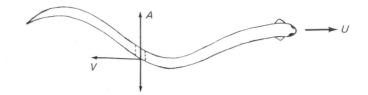

Fig. 6.2. A generic undulator swims at forward speed U by propagating waves of body bends at a speed V and amplitude A. This figure applies to either high or low Reynolds number undulators.

μm long to eels reaching 1 m in length. While a superficial view suggests that mechanism underlying thrust generation is similar, the physics of flows over such size ranges suggests otherwise. We use the Reynolds number ($R_e = Ul/\nu$, where U is the forward speed of the animal; l, its length; and ν, the kinematic viscosity of the fluid) as an indicator of the different physical realms associated with large size and speed ranges. When R_e is very low ($<< 1$), viscous forces dominate the motion, inertia is negligible and thrust depends largely on the relative speed of undulating body segments. As an example, the Reynolds number for sperms is about 10^{-3} and their thrust is proportional to the speed of the bending waves that propagate along their flagellum. Eels, at the other extreme, have body speed and size that gives a Reynolds number of about 10^5. At this range of Reynolds numbers, thrust arises from the lateral accelerations of body segments.

Do thrust and efficiency have different behaviors for these extremes in Reynolds number? Simple theoretical estimates of these parameters provide some insight into this question. Consider a generic undulator which moves forward at speed U that has a wave of amplitude A propagating rearward with speed V with a lateral frequency ω (Fig. 6.2). For organisms undulating at the two extremes in Reynolds number using the above kinematics, a comparison of equations for thrust and efficiency show several intriguing trends (Table 6.1). First, increases in the rearward wave speed (V) increase thrust as the square of that parameter in high Reynolds number and linearly at low Reynolds numbers. Thrust also increases with the frequency of undulations and, generally, with their amplitude. Intriguingly, the efficiency follows a quite different behavior: (1) it decreases similarly at both Reynolds number extremes as the relative rearward wave speed increases; and (2) it is independent of either the frequency or amplitude of the wave speed.

Table 6.1. *Expressions for thrust and efficiency in steady state aquatic locomotion.*

Mode		Thrust	Efficiency
Undulation (High Re)		$AF^2[1 - (U/V)^2]$	$1 + (U/V)/2$
Undulation (Low Re)	$VL[1 - (U/V)][(1 + A/FV)^{-1} - 1]$		$U(1 - K_T/K_N)/V$
Lift (High Re)		$(FA)^2(1 + e^{2FL/U})/2$	$(1 + e^{-FL/U}))/2$
Drag (Any Re)*		$SK_N U^2[1 - (U/V)^2]$	0.1–0.8

Parameters are: A = amplitude of motion, f = frequency of motion,
U = forward speed of animal, V = rearward wave speed,
K = drag coefficients at high or low Reynolds number with subscripts T defining
tangential flow and N defining normal flows,
L = length of animal, and S = area of propulsor.
Equations are modified from Lighthill (1975), Childress (1981),
Daniel (1988) and Blake (1983). Efficiency estimates for rowers come from
Nachtigall (1960) and Blake (1981a).
An asterisk denotes maximum estimates.

In these examples, efficiency is at least limited by the need for movement, reaching a maximum of 1.0 only when thrust is zero! However, for any set value for efficiency (less than 1) thrust may be increased by changes in the amplitude or frequency of the propulsive wave. At first glance it appears that thrust may rise indefinitely with fixed values of efficiency. In reality, however, the theories underlying the equations in Table 6.1 neglect any physiological limits imposed on the total power output of the organism and they neglect the complicated flows that arise when organism motions become so large that the two-dimensional assumptions of the theories are violated. Both of these factors can potentially limit the thrust and, therefore, the efficiency. Neither has been fully explored in studies of aquatic locomotion.

6.3.2 Locomotor efficiency in lift-based propulsion

For many aquatic organisms such as rays, skates, tuna, penguins, and whales thrust is generated by a scheme quite different from the examples above. Rather than relying on forces that act parallel to the direction of

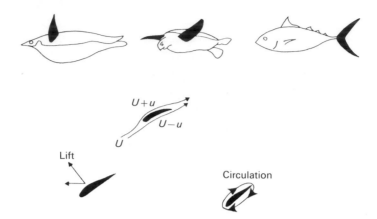

Fig. 6.3. The diversity of aquatic flyers is shown with a resolution of the instantaneous lift vector into a thrust (forward) and induced drag (rearward) component. At the lower right portion, the circulatory flows set up by wing-like appendages is shown in a frame of reference moving with the wing at speed U.

flow relative to body segments, the propulsive appendages in these animals generate lift forces that act perpendicular to the direction of motion fluid motion (Fig. 6.3). This lift-based mechanism is restricted only to organisms inhabiting the realm of high Reynolds numbers in which inertia plays a prominent role. A key to understanding the mechanism of lift production is seen in Fig. 6.3 in which, for a reference frame moving a wing speed U, there is a circulation of fluid clockwise around the wing. The magnitude of the lift is proportional to the level of circulation about the wing, and increases as the square of the wing speed. This lift vector, acting normal to the surface of the wing, has a component in the thrust direction.

For most aquatic flyers, the lift depends not just on the instantaneous speed, but also on the frequency with which the wing oscillates. This dependence arises from the fact that circulation takes time to develop, and in oscillatory motions changes in circulation do not quite keep up with changes in wing direction (Lighthill, 1975; Wu, 1971; Daniel, 1988). The reduced frequency parameter (σ) measures the amount of this oscillatory motion that contaminates an otherwise steady-state view of the wing. It is simply the ratio of the oscillation speed divided by the forward speed ($\sigma = \omega/U$ where l is the dimension of the wing parallel to fluid motion). Theoretical results show that both the thrust generally rises with frequency and the efficiency decays exponentially with that parameter (Table 6.1). As we saw above in the examples of undulators, thrust and efficiency are inversely related and, again, efficiency is independent of the amplitude of wing motion. These theories also assume reasonably small amplitudes and neglect any physiological limits to the total power output.

6.3.3 Locomotor efficiency in rowing propulsion

Aquatic beetles, labriform fishes, frogs and a variety of planktonic crustaceans all propel themselves with rowing motions of oar-like appendages. They generate thrust with a combination of resistive and reactive forces that depend on fluid speed and acceleration respectively (Table 6.1). The extent to which each of these mechanisms contribute to thrust depends on the Reynolds number of the appendage. In angelfish, for example, reactive forces predominate in thrust production (Blake 1981a,b) whereas in small aquatic beetles, resistive forces alone give reasonable estimates of the total thrust (Nachtigall, 1960; Nachtigall and Bilo, 1975). In both cases, increases in the amplitude and frequency of appendage motion yield greater thrust.

While simple theoretical expressions for the efficiency of rowing locomotion are not available, Nachtigall's (1960) data gives a Froude efficiency of 0.8 for beetle hindlimbs. Blake (1981a,b), however, shows that the efficiency of angelfish fins may be as low as 0.12 if one accounts for the energy required to accelerate the mass of the limb and surrounding water. Neither of these estimates apparently depends upon the amplitude of the appendage motion.

6.3.4 General results of steady-state swimming

Despite differences in the mechanisms associated with producing thrust, two general points emerge: (1) thrust and efficiency are inversely related; and (2) efficiency does not depend upon the amplitude of appendage motion and, in some cases, is independent of the frequency of motion. The consequences of these results are that for fixed levels of efficiency, theory predicts no limits to thrust. Under a limit to the maximum power output of muscle, the range of motions that are physiologically feasible limits the range of thrust and efficiency (Daniel and Webb, 1987). Below is a case study that examines how such factors may influence the levels of thrust and efficiency.

While it is not a subject here, we must also realize that as the amplitudes of the motions become very large, many of the assumptions underlying the theories are violated and we should see decrements in the expected total thrust with such large-amplitude motions.

6.4   EFFICIENCY IN UNSTEADY SWIMMING

The above examples all point to some common limits to the efficiency of aquatic organisms that move at constant velocity. In many biological situations, however, constant forward speed rarely occurs. Instead, turns and accelerations may dominate the kinematics. For fish escape and attack maneuvers (Weihs, 1972; Webb, 1979,1983; Harper and Blake, 1989) and invertebrate examples such as squid (O'Dor, 1988), shrimp (Daniel and Meyhofer, 1989), and chaetognaths (Jordan, 1990) acceleration is seen as a key measure of performance. How might efficiency be formulated for such circumstances? One possible measure is the ratio of the rate at which energy is expended in accelerating the body to the maximum available power output. For this measure, one would expect the efficiency to be highest if most of the available power goes to doing useful work on the environment rather than to tissue deformation.

6.4.1   Rotations and translations

Fish and shrimp both undergo very stereotypical escape motions in response to attack, and the mechanics and kinematics of these two examples have been studied in some detail (Weihs, 1972; Webb, 1983; Harper and Blake, 1989; Daniel and Meyhofer, 1989). In fish, escape movements are produced by a series of body flexions that begin with a single bend of the animal into a C-shape, followed by a rapid, sinuous, flexion of the body in the opposite direction (Weihs, 1972). In shrimp, a powerful flexion of the abdomen propels the animal tailward in the classic tail-flip response. In both animals, the kinematics of body movements lead to an asymmetric thrust production, giving rise to both translational and rotational accelerations of the body. As is the case for steady-state swimmers, large amplitude motions, greater rates of movement, and larger propulsive structures all lead to greater total forces (Weihs, 1972; Daniel and Meyhofer, 1989). However, because both rotation and translation result from asymmetric forces about the center of the animal's mass, increases in total force do not necessarily lead to increases in performance as measured by linear acceleration, distance travelled, or total useful work.

In recent study of shrimp escape maneuvers (Daniel and Meyhofer, 1989) a theoretical analysis which balances both linear and angular momenta for these animals finds that increases in the relative size of the abdomen do not give rise to increases in performance. Indeed, there is a unique relative abdomen size that maximizes performance. This maximum follows from the fact that extremely small abdomens (less than 50% of the body length) yield very small forces which barely translate or rotate the body. As the abdomen increases in size, thrust forces increase, but the moment arm of forces about the center of gravity increase as well. Eventually, the moment arm may become so large that the dominant body kinematics will be rotational rather than translational. Performance in this case, and in the case of C-starts in fish, is limited by the need for translation. In both cases, we predict morphologies that maximize the useful thrust.

6.4.2 Maximum muscle stress

Unfortunately, in the above analyses, as well as those for the steady-state examples, physiological limitations to the maximum forces remained unexplored. In reality, propulsive muscles generating such forces may have very severe limits, one of which is the maximum isometric stress (force per unit cross-sectional area of muscle) they can produce. Indeed, this maximum stress appears to be quite constant for many muscle types. In vertebrate striated muscle, for example, about 100 kPa is quite common (McMahon, 1984), whereas as in rapid flexors of shrimp, a value 10 kPa is found (Atwood, 1973). This limitation imposes a constraint on the range of appendage or body morphologies and kinematics that can lead to increases in thrust, even under fixed levels of efficiency. For example, speeds and amplitudes of body movements may only increase to the point where the internal stresses required to produce thrust forces do not exceed the above maximum values. The magnitude of such stresses been examined only in terrestrial systems (Biewener et al., 1988, and see Chapter 5).

To examine performance in light of this physiological limit, theoretical models of the tail-flip in shrimp were developed to constrain the motion of the abdomen so that the maximum stress is constant throughout the maneuver (Daniel and Meyhofer, 1989). This model was used to examine how escape performance with physiological limits scales across a broad size range of shrimp, The results are that (i) as the overall size of the animal increases, tail motions must become ever slower to conserve the muscle stress, and (ii) there is a unique size and shape that maximizes escape performance. The maximum follows from the fact that at very small sizes, tail motions can be quite rapid and rotational moments become quite large. At the other extreme in size, tail motions become so slow as to produce too little thrust for accelerating the mass of the animal.

The imposition of a physiological constraint to body kinematics provides a first glimpse into limits to performance in aquatic animals. It is important to realize, however, that additional physiological factors may come into play. For example, there is an important inverse relationship between the total force developed by propulsive muscles and the speed with which they contract (McMahon, 1984). This relationship, known as Hill's equation, has a maximum in the total power output of muscles at some particular speed of contraction. With these features of the propulsive muscles, the range of kinematics and morphologies that are physiologically feasible are still further constrained. At present, there is no study that incorporates both the hydrodynamics of locomotion and the mechanics of force development in contracting muscle to examine limits to overall thrust and efficiency.

6.4.3   *Viscosity and elasticity of muscle cells*

In addition, to the concepts developed above, other mechanical factors in locomotor systems may constrain the magnitude of useful work done, even within the framework of physiologically feasible body motions. For example, according to Fig. 6.1, the amount of energy required to deform tissues, including muscles, will determine the overall efficiency of locomotion. If this energy is large, and the motion is a simple flip, then the fraction of the total available power going to propulsion is reduced. In repeated motions, however, some of that elastic energy is potentially recoverable and, in terrestrial systems, this energy storage may yield a net decrease in the overall energy requirements for locomotion (Alexander *et al.*, 1982; Cavagna *et al.*, 1977). In either case, we would hope that the amount of mechanical energy that is dissipated in the muscle fibers and surrounding tissues would be small.

Measurements of the energy storage and dissipation in both vertebrate and invertebrate muscle fibers reveal some rather disturbing trends in the amount of energy they dissipate (Tidball and Daniel, 1986; Meyhofer and Daniel, 1990). Over a broad range of deformation frequencies and sarcomere lengths, extensor muscle cells from the abdomen of shrimp dissipate up to 50% of the energy imparted to them (Meyhofer and Daniel, 1990). Moreover, this peak in energy dissipation occurs in the middle of the physiological range of frequencies and lengths. More intriguingly, a simple calculation shows that the energy dissipated in both flexor and extensor cells during the rapid tail flip is about equal to the total energy required to accelerate the mass of the animal (Meyhofer and Daniel, 1990)! This rather large dissipation of energy begs an important question: If overall efficiency is important, why does the viscous dissipation of energy require such a significant fraction of the total available energy?

One answer to that question is that biological materials are inherently inefficient. Unfortunately, this pessimistic view is not supported by available data on muscle or other biological materials (see Gosline, Chapter 4). Indeed, outside of the physiological range of deformation frequencies and sarcomere lengths, muscle appears to be quite elastic (Tidball and Daniel, 1986; Alexander and Bennet-Clark, 1977; Cavagna *et al.*, 1977). An alternative view is that energy dissipation may not be 'bad' and the some energy must be dissipated in order to respond stably to impulsive forces. Indeed, any system that has mass and elasticity can respond to dynamic loads with ringing, undamped oscillations, or the production of locally high internal strains (Timoshenko *et al.*, 1974). In rapid locomotor movements, such responses could potentially damage muscle cells or other tissues. Viscous energy dissipation is a mechanism that can prevent such problems from arising. Thus, at the level of muscle fibers associated with locomotion, there is a limit to the maximum efficiency that is set by the need for stability. The implications of this need for dissipation remain unexplored in nearly all studies of animal locomotion.

6.5   SUMMARY

This Chapter examined the thrust and efficiency for aquatic locomotion in light of two basic differences in the type of movement. For steady-state swimmers, those with constant velocity, theories show that thrust and efficiency are inversely related despite a broad range of modes of propulsion, swimming speeds, and animal sizes. Moreover, a general result arises that, for fixed levels of efficiency, thrust may be altered by changes in body morphology or the kinematics of the thrust-producing portions of the body. Unfortunately, these theories provide no insight into physiological limits to thrust and, therefore, limits to the efficiency that can be obtained for given values of thrust. A brief tour of unsteady swimmers – those that accelerate to attack their prey or avoid predation – shows that important mechanical limits to thrust arise when forces are generated asymmetrically about an animal's center of mass. More detailed views of these unsteady modes of movement show that physiological limits to the forces and speeds of contracting muscle impose constraints on the range of feasible motions and morphologies. At even lower levels of biological organization, we find that the mechanics of the cells involved in producing thrust impose still further constraint on thrust and body kinematics. These latter constraints arise from the need for energy dissipation in muscle cells, a characteristic that yields stability in the response of these cells to the dynamic loads present in locomotor movements

6.6 REFERENCES

Alexander, R.McN. and Bennet-Clark, H.C. (1977). Storage of elastic strain energy in muscle and other tissues. *Nature, Lond.* **265**:114–17.

Alexander, R.McN., Maloiy, G.M.O., Ker, R.F., Jayes, A.S. and Warui, C.N. (1982). The role of tendon elasticity in the locomotion of the camel (*Camelus dromedarius*). *J. Zool., Lond.* **198**:293-313.

Atwood, H.L. (1973). An attempt to account for the diversity of crustacean muscle. *Am. Zool.* **13**:357–78.

Biewener, A.A., Alexander, R. McN. and Heglund, N.D. (1981). Elastic energy storage in the hopping kangaroo rats (*Dipodomys spectabilis*). *J. Zool., Lond.* **195**:369–83.

Biewener, A.A., Blickhan, R., Perry, A.K., Heglund, N.D. and Taylor, C.R. (1988). Muscle forces during locomotion in kangaroo rats: force platform and tendon buckle measurements compared. *J. exp. Biol.* **137**:191–205.

Blake, R.W. (1981a). Mechanics of drag-based mechanisms in propulsion of aquatic vertebrates. *Symp. Zool. Soc. Lond.* **48**:29–52.

Blake, R.W. (1981b). Influence of pectoral fin shape on thrust and drag in labriform locomotion. *J. Zool., Lond.* **195**:53–66.

Blake, R.W. (1983). *Fish Locomotion.* London: Cambridge University Press.

Childress, S. (1981). *Mechanics of Swimming and Flying.* New York: Cambridge University Press.

Cavagna, G.A., Heglund, N.C. and Taylor C.R. (1977). Mechanical work in terrestrial locomotion: two basic mechanisms for minimizing energy expenditure. *Am. J. Physiol.* **233**:243–61.

Daniel, T.L. (1988). Forward flapping flight from flexin fins. *Can. J. Zool.* **66**:630–8.

Daniel, T.L. and Webb, P.W. (1987). Physical determinants of locomotion. In *Comparative Physiology: Life in Water and on Land* (ed. P. Dejours, L. Bolis, C.R. Taylor and E.R. Weibel), pp. 343-69. Padova: Liviana Press.

Daniel, T.L. and Meyhofer, E. (1989). Size limits in escape locomotion of carridean shrimp. *J. exp. Biol.* **143**:245–65.

Jordan, C.E. (1990). Swimming behaviour of planktonic predators: undulatory locomotion of the chaetognath *Sagitta elegans. J. exp. Biol.* (submitted).

Lighthill, M.J. (1975). *Mathematical Biofluiddynamics.* Philadelphia: Soc. Ind. App. Math.

McMahon, T.A. (1984). *Muscles, Reflexes and Locomotion.* New Jersey: Princeton University Press.

Meyhofer, E. and Daniel, T.L. (1990). Dynamic mechanical properties of extensor muscle cells of the shrimp *Pandalus danae:* cell design for escape locomotion. *J. exp. Biol.* **151**, 435–52.

Nachtigall, W. (1960). Uber kinematic, dynamic und energetik des schimmens einheimischer Dyytisicen. *Z. Vergl. Physiol.* **43**:48–180.

Nachtigall, W. and Bilo, D. (1975). Hydrodynamics of the body of *Dytiscus marginalis* (*Dytiscidae, Coleoptera*). In *Swimming and Flying in Nature* (ed. T.Y. Wu, C.J. Brokaw, and C. Brennen), pp. 585–95. New York: Plenum Press.

O'Dor, R.K. (1988). The forces acting on a swimming squid. *J. exp. Biol.* **137**:421–42.

Tidball, J.G. and Daniel, T.L. (1986). Elastic energy storage in rigored skeletal muscle cells under physiological loading conditions. *Am. J. Physiol.* **250**, R56–R64.

Timoshenko, S., Young, D.H. and Weaver, W. (1974). *Vibration Problems in Engineering.* New York: John Wiley.

Webb, P.W. (1979). Mechanics of escape response in crayfish (*Oronectes virilis* Hagen). *J. exp. Biol.* **102**:245–63.

Webb, P.W. (1983). Speed, acceleration and manoeuvrability of two teleost fishes. *J. exp. Biol.* **102**:115–22.

Webb, P.W. and Weihs, D. (1983). *Fish Biomechanics.* New York: Pracger Press.

Weihs, D. (1972). A hydrodymamic analyisis of fish turning manoeuvres. *Proc. R. Soc. B* **182**:59–72.

Weihs, D. and Webb, P.W. (1983). Optimization of locomotion. In *Fish Biomechanics* (ed. P.W. Webb and D. Weihs), pp. 339-71. New York: Preager Press.

Wu, T.Y. (1971). Hydromechanics of swimming propulsion. Part 2. Some optimum shape problems. *J. Fluid Mech.* **46**:521–44.

Wu, T.Y. (1975). Hydromechanics of fish swimming. In *Swimming and Flying in Nature* (ed. T.Y. Wu, C.J. Brokaw, and C. Brennen), pp. 615–34. New York: Plenum Press.

7

The concepts of efficiency and economy in land locomotion

R. J. FULL

7.1 INTRODUCTION

Is the concept of efficiency in terrestrial locomotion useful when applied at the level of the whole animal? Can measures of efficiency aid in revealing the consequences of variation in morphology and physiology that relate to locomotion on land? Can these measures quantify the effect of variation in leg number (two in humans to over 600 in a milli-pede), leg length and orientation, stepping pattern (metachronal waves gaits, trotting versus hopping), muscle type, musculo-skeletal arrange-ment (exo- versus endoskeletons), body shape (long in centipedes and millipedes versus round in some crabs), and locomotor style (forwards versus sideways travel in crabs)? Can measures of efficiency provide use-ful information about the mechanistic, ecological and evolutionary bases of how animals of diverse body form move? Most lay people, as well as researchers, would probably answer yes to each question. Many func-tional morphologists, physiologists and biomechanists believe they can recognize efficient terrestrial locomotion. Unfortunately, when our hy-potheses are tested, the results are often ambiguous. Whether or not our hypotheses are based on sound principles of physics or physiology seems to make little difference. Awkward marathon runners win gold medals; athletes ranked as inefficient expend less energy than more graceful run-ners (Cavagna and Kram, 1985); some mammals with large limbs use no more energy than those that evolved more 'efficient' tapered limbs (Taylor et al., 1974); and animals, such as a centipede, that 'waste' mo-tion as they laterally undulate actually require somewhat less energy to travel a given distance than other animals of the same mass (Full, 1989).

7.2   WHAT IS THE DEFINITION OF EFFICIENCY ?

How can we resolve this apparent mismatch between reasonable hypotheses and evidence? There are at least two reasons for the mismatch. One has its origins in the definition and use of the term efficiency, whereas the other results from the complexity of the systems used in locomotion. What is the definition of the term efficiency and how do we apply it to comparative terrestrial locomotion? Efficiency has at least two definitions and is most often used in two ways. Efficiency, in the context of mechanics, is defined as:

(n.) 1. the ratio of work or energy (E) output to input
(mechanical efficiency $= E_{output}/E_{input}$).

This definition most often refers to animals performing straight-ahead, constant-speed locomotion. Measurements of both input and output have been conducted on surprisingly few species (Alexander and Vernon, 1975; Cavagna *et al.*, 1977; Dawson and Taylor, 1973; Blickhan and Full, 1987; Heglund *et al.*, 1982b; Herreid and Full, 1984; Full, 1987; Full and Tu, 1989). Mechanical efficiency simply relates one measure of performance, metabolic energy input, to one possible explanation of its variation, mechanical energy output.

A second definition of efficiency is:

(n.) 2. effectiveness or competency in performance.

It is often this second notion of effective performance that best describes the common use of the term efficiency. This definition has considerable utility in addressing mechanistic, ecological and evolutionary questions.

7.3   WHAT IS THE 'EVENT' AND HOW DO YOU MEASURE PERFORMANCE?

When we make a statement concerning an animal's efficiency, two questions must be answered. First, exactly what task or 'event' is the animal performing? Second, what is the measure of performance? In many cases, our hypotheses concerning the efficiency of one animal compared to another would probably be supported, but we simply may not be conducting the appropriate test, staging the appropriate 'event' to measure performance. For example, an animal may be very inefficient at steady-state, constant-speed locomotion, but may be highly effective at maneuvering around or over obstacles. Likewise, animals that are very effective sprinters may be inefficient steady-state runners when migrating long distances. Selecting the 'event' to study for comparison is crucial and cannot be ignored. To best study the consequences of variation in locomotor structure and function (such as leg number, leg position, skeletal type), it is obvious that efficiency or effectiveness indices must include not only steady-state locomotion, but also intermittent activity, obstacle negotiation (e.g. climbing, leaping and maneuvering) and burst locomotion.

Selecting the relevant measures of performance may be equally as important as selecting the appropriate event. Inappropriate measures of performance could result in a mismatch between data and our hypotheses of efficiency. What are the criteria used to evaluate locomotion? If we compare the performance of animals moving over a comparable distance, does the efficient animal use the least energy, travel the distance in the shortest time, or recover the fastest for the next bout of activity? Energy utilization is not the only measure of effective performance. Obviously, we demand other performance measures when purchasing wheeled vehicles, such as automobiles, motorcycles and trucks. Performance measures other than energy utilization are receiving more attention in legged locomotion for a variety of species. Some of these include: (1) endurance or fatigue resistance, (2) distance travelled, (3) acceleration and maximum speed, (4) durability and strength, and (5) maneuverability and stability.

7.4 QUESTIONS ADDRESSED BY COMPARING PERFORMANCE

Variation in locomotor performance within and among species has and will continue to reveal clues to the mechanistic bases of muscular, skeletal and nervous systems. Considering multiple events and performance measures is likely to reveal interesting functional and structural compromises, since few complexes are optimized for a single event. For example, Biewener and Blickhan (1988) found that, for their mass, kangaroo rats have proportionately large hind limb muscle, tendons and bones to withstand the large forces associated with rapid acceleration during predator avoidance. These relatively large structures, however, limit their ability to store and recover elastic strain energy during steady-state locomotion. Using the concept of efficiency as effective performance cannot

only address mechanistic questions, but also permits the formulation
of evolutionary and ecological hypotheses. Recent attempts have been
made to explain variation in locomotor performance in terms of ecology
and evolution (Huey, 1987; Huey and Bennett, 1986). The evolution
or origin of morphological and physiological characters is studied by
mapping variation in performance onto independently established phylogenies. For example, evolutionary change in relative hind limb length
of Anolis lizards can explain much of the variation measured in sprint-
ing and jumping performance (Losos, unpublished). Maintenance and
selection of locomotor characters is addressed in an ecological context by
measuring variation in performance within populations (Arnold, 1983).

7.5   METABOLIC ENERGY INPUT - THE ECONOMY
OF LOCOMOTION

The economy of locomotion is one performance measure that has been
used frequently to evaluate efficiency in both senses of the word, effectiveness and mechanical efficiency. The economy of locomotion, metabolic energy input, is the denominator in the mechanical efficiency
equation. It is commonly represented by the submaximal, steady-state
oxygen consumption (V_{O2ss}) per unit time (time-specific economy) or
per unit distance (distance-specific economy) of an animal running at
a constant speed on a treadmill. Animals considered to be economical
have lower values of submaximal, steady-state oxygen consumption per
unit time or distance. Can the economy of locomotion, as a measure of
efficiency, aid in revealing the consequences of variation in morphology
and physiology? The answer depends upon whether: (1) we can obtain
comparable measures of economy in diverse species, (2) significant variation in economy exists, and (3) variation in economy can be related to
variation in morphology and physiology.

7.5.1   A. What is the common currency?- Is oxygen consumption the appropriate measure?

For most legged runners, steady-state oxygen consumption (V_{O2ss})
during constant-speed treadmill exercise is a reasonable indicator of the
energy used. Non-aerobic contributions appear to be negligible at V_{O_2}
rates below 80% of maximum. The oxygen transport systems of birds
and mammals (Seeherman *et al.*, 1981), lizards (Seeherman *et al.*, 1983),

insects (Herreid and Full, 1984), centipedes and millipedes (Full and Herreid, 1986), some crustaceans (Full, 1987; Full and Herreid, 1983; Herreid and Full, 1985) and even a lungless salamander (Full, 1986) adequately delivery oxygen to aerobically functioning muscles. However, the assumption that V_{O2ss} is a reliable measure of metabolic power input must be checked, since it is violated in other species, such as some lungless salamanders, crabs and spiders, in which V_{O2} uptake kinetics are very slow (half time to steady-state is 1 6 min) and/or locomotion at even very slow speeds represents a large fraction of V_{O2max} (Full and Herreid, 1984; Full *et al.*, 1985; Herreid *et al.*, 1983; Anderson and Prestwich, 1985; Prestwich, 1983). In the future, more studies of anaerobic end-product (e.g. lactate) production and removal are needed to better quantify the contributions of accelerated glycolysis or other anaerobic pathways (Brooks *et al.*, 1984).

7.5.2   B. Where is metabolic energy used during locomotion?

Certainly, most of the metabolic energy used during locomotion is required by contracting locomotor muscles (Fig. 7.1). Yet, other tissues and organs must be maintained. Some may demand less energy during activity, such as digestive and excretory processes (Stainsby *et al.*, 1980). On the other hand, hormone release during exercise can elevate tissue respiration during and after exercise (Cain, 1971). An increased energy demand may result from a Q_{10} effect on tissues as body temperature rises or additional energy may be required for thermoregulation (Hurst *et al.*, 1982). The cost of operating respiratory and circulatory systems at higher rates during exercise may vary among species (see Milsom in Chapter 8). Are circulatory costs lower or higher in open versus closed systems? Is the metabolic cost of gill ventilation in crabs substantial and does it increase with speed? How does this cost compare to the cost of lung ventilation in lizards or tracheal ventilation in insects as a function of speed? Do lungless salamanders save energy during exercise by having no ventilatory costs? One important issue that must be addressed is whether or not whole animal measures of oxygen consumption can allow partitioning of these different sources demanding energy.

7.5.3   C. Gross, net and incremental economy - which to use for efficiency?

In the majority of animals tested thus far, including, birds and mammals (Taylor *et al.*, 1970; Seeherman *et al.*, 1981), lizards (Bennett,

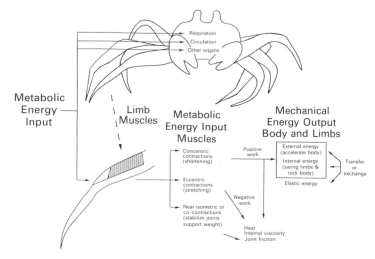

Fig. 7.1. Metabolic energy input and output in land locomotion. Metabolic energy is used by both locomotor muscles and non-locomotor sources. Limb muscles require energy when they shorten (concentric contraction), are stretched (eccentric contraction), or used to stabilize joints and support an animal's weight. Metabolic energy is dissipated due to friction and lost as heat. Concentric contractions generate increases in mechanical energy to move the body (external energy) and limbs and trunk relative to the body (internal energy). Eccentric contractions are associated with decreases in mechanical energy. Mechanical energy can be transferred from one source to another and between segments, as well as within a segment.

1982; Seeherman *et al.*, 1983), insects (Herreid *et al.*, 1981), centipedes and millipedes (Full and Herreid, 1986), some crustaceans (Full, 1987; Full and Herreid, 1983; Herreid and Full, 1985), and lungless salamanders (Full, 1986), V_{O2ss} or, better, the gross rate of oxygen consumption ($V_{O2gross}$) increases linearly with speed below the maximum oxygen consumption. Ponies forced to walk or trot at higher or lower than normal speeds (Hoyt and Taylor, 1981), walking humans (Margaria *et al.*, 1963), and running polar bears(Hurst *et al.*, 1982), squirrels (Hoyt and Kenagy, 1988), and minks (Williams, 1983) are exceptions, showing curvilinear or discontinuous functions. Hopping kangaroos decrease $V_{O2gross}$ with speed (Dawson and Taylor, 1973).

Traditionally, the linear $V_{O2gross}$ vs. speed function has been partitioned into three components, maintenance (V_{O2mat}), offset ($V_{O2offset}$) and incremental oxygen consumption (V_{O2inc}; Fig. 7.2) where:

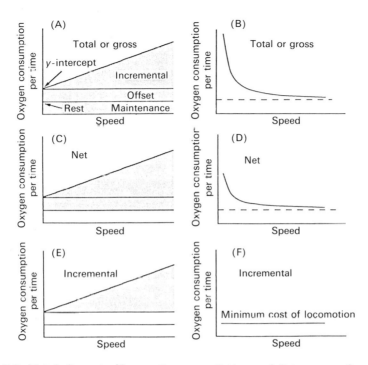

Fig. 7.2. Metabolic cost of locomotion per unit time and distance as a function of speed. (A) Metabolic energy per time (rate of oxygen consumption) increases linearly with speed. Total or gross metabolic cost (shaded area) is the sum of three components; rest, offset and incremental. (B) Total metabolic cost per distance. Dashed line represents the minimum cost of locomotion. (C) Net metabolic cost per time (shaded area) includes offset and incremental costs. (D) Net metabolic cost per distance. Dashed line represents the minimum cost of locomotion (E) Incremental or work metabolic cost per time. (F) Incremental metabolic cost per distance or the minimum cost of locomotion which equals the slope of the metabolic cost versus speed function.

$$V_{O2ss} = V_{O2gross} = V_{O2mat} + V_{O2offset} + V_{O2inc} \qquad (7.1)$$

This partitioning of steady-state oxygen consumption has lead to at least three different economy values, gross ($V_{O2gross}$), net (V_{O2net}) and incremental (V_{O2inc}). These values have been used extensively in human and animal exercise physiology. Disagreement exists over which value is the most appropriate measurement of economy (Donovan and Brooks, 1977; Stainsby *et al.*, 1980).

(1) **Gross or total metabolic cost** is the sum of all three compo-

nents of the V_{O_2} vs. speed function, $V_{O2offset}$, V_{O2mat} and V_{O2inc} (Fig. 7.2A). Time-specific, gross metabolic cost increases with speed, whereas distance-specific, gross metabolic cost decreases with speed (Fig.2B). (2) **Net metabolic cost** equals the gross metabolic cost minus resting oxygen consumption (V_{O2rest}) and assumes that V_{O2rest} equals V_{O2mat} at all speeds (Fig. 7.2C). Time-specific, net metabolic cost increases with speed, whereas distance-specific, net metabolic cost decreases with speed (Fig. 7.2D). Net metabolic cost includes $V_{O2offset}$. (3) **Delta, incremental or instantaneous metabolic cost** equals the gross metabolic cost minus V_{O2mat} and $V_{O2offset}$ (Fig. 7.2E). Time-specific, incremental metabolic cost increases with speed, whereas distance-specific, net metabolic cost is independent of speed (Fig. 7.2F). Variation in incremental metabolic cost can result if $V_{O2offset}$ or V_{O2mat} are not baselines that remain constant as a function of speed.

7.5.4   D. Maintenance and 'offset' costs during locomotion - What are the bases of baselines and are they lines?

(1) Maintenance costs   Variation in V_{O2mat} can significantly affect metabolic cost if $V_{O2gross}$ is used. Direct measures of V_{O2mat} are difficult and few estimates have been attempted. In most cases a constant baseline, approximated by V_{O2rest}, is the standard assumption (Fig. 7.2A; Stainsby *et al.*, 1980). This assumption needs further testing, especially if maintenance components are large fractions of the total or gross $V_{O2gross}$ and increase or decrease as a function of speed.

Several variables such as body mass, temperature and species affect V_{O2rest}.

Body mass. Much of the variation in V_{O2mat} can be attributed simply to body mass. Mass-specific V_{O2rest} decreases with increasing body mass in nearly all animals; however scaling exponents show considerable variation (Fig. 7.3; Hemmingsen, 1960).

Temperature. In most ectotherms, changes in temperature will add to this variation. V_{O2rest} approximately doubles with a $10^{o}C$ increase in temperature, but Q_{10} values also vary.

Body form – species differences. Considerable variation in maintenance cost is present for different species or taxa of the same body mass measured at the same temperature. Surprisingly, ectotherms require about only one-quarter the V_{O2rest} of an endotherm, even when both are operating at comparable body temperatures (Bennett, 1982). Considerable variation also exists among ectotherms. For example, sala-

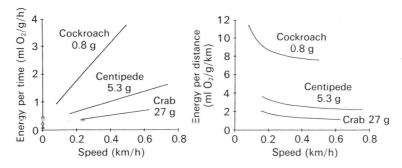

Fig. 7.3. Mass-specific oxygen consumption per time and distance as a function of speed for animals that differ in body mass and form. (A) Despite variation in form mass-specific maintenance, offset and incremental oxygen consumption all increase with a decrease in body mass. Resting oxygen consumption is shown for the cockroach (square), centipede (circle), and crab (triangle). (B) Smaller animals require more metabolic energy to move a gram of mass one meter. Data on ghost crabs from Full (1987). Cockroach and centipede from Full unpublished.

manders have an unusually low V_{O2rest} compared to other ectotherms at the same temperature (Fig. 7.4; Full *et al.*, 1988; Walton, unpublished).

(2) 'Offset' cost Animals may differ in gross or net cost of locomotion because of changes in 'offset' costs. $V_{O2offset}$ is energy used in addition to maintenance requirements at near zero speed. It results from the fact that the V_{O2ss} vs. speed function extrapolates above resting oxygen consumption V_{O2rest} at zero speed. It is calculated by subtracting V_{O2rest} from the *y*-intercept. Just as for maintenance costs, a constant baseline with an increase in speed is also the standard assumption for offset costs (Fig. 7.2B). Postural costs, stress, and an elevated body temperature have all been proposed as explanations of $V_{O2offset}$, including the possibility that it is an artifact resulting from the lack of data at very low speeds (Herreid, 1981; Herreid and Full, 1988; Schmidt-Nielsen, 1972). Surprisingly, hermit crabs without their shells do not show a $V_{O2offset}$ component; the V_{O2ss} vs. speed relationship extrapolates to V_{O2rest} (Herreid and Full, 1985). Locomotion at even the slowest speeds requires additional energy $V_{O2offset}$ to carry a shell that is about equal in weight to their own body. Perhaps $V_{O2offset}$ does correspond to the energy cost initially required to lift the center of mass. At present $V_{O2offset}$ has no identifiable physiological basis.

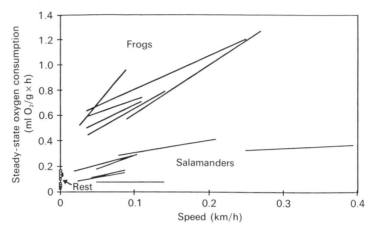

Fig. 7.4. Mass-specific oxygen consumption for frogs and salamanders of similar mass measured at comparable temperatures. Walking salamanders have significantly lower resting, offset and incremental metabolic costs. (Gatton, Miller and Full, in press). Resting oxygen consumption is shown by circles for frogs and squares for salamanders.

As with V_{O2mat}, $V_{O2offset}$ can be affected by body mass, temperature and species differences.

Body mass. $V_{O2offset}$ ranges from 30 to 190% of V_{O2rest} in birds and mammals, but averages approximately 70% of V_{O2rest} (Palidino and King, 1979). Mass-specific $V_{O2offset}$ decreases with increasing body mass in birds and mammals (Palidino and King, 1979, Taylor *et al.*, 1982). A similar relationship with body mass is likely in other animals, such arthropods (Fig. 7.3) and lizards. However, the y-intercept is more variable in these groups, because few animals have actually been measured at comparable temperatures.

Temperature. Increased temperature elevates $V_{O2offset}$ in ectotherms, but does not affect incremental cost (Herreid *et al.*, 1981). Some cockroaches and lizards double $V_{O2offset}$ with a 10°C increase in temperature, whereas others shower smaller Q_{10} effects (Herreid *et al.*, 1981; John-Alder and Bennett, 1981).

Body form - species differences. 'Offset' costs have been shown to vary considerably among species when the effects of body mass are removed. Salamanders have a low $V_{O2offset}$ compared to frogs even when measured at comparable temperatures (Fig. 7.4; Full *et al.*, 1988; Walton, unpublished). The y-intercept of running minks is 56% higher than predicted and cannot be accounted for by an elevated V_{O2rest} (Williams, 1983). Intraspecific variation is also apparent in $V_{O2offset}$. The mass-specific $V_{O2offset}$ of ghost crabs does not decrease with an increase in body mass (Full, 1987). At low speeds larger crabs actually require more energy on a per gram basis to locomote due to a high $V_{O2offset}$.

(3) Incremental or delta costs - the minimum cost of transport   At least at high speeds, incremental or delta economy probably best represents locomotor muscle costs, because V_{O2mat} and $V_{O2offset}$ are likely to become small percentages of the gross cost (Fig. 7.2E). Margaria (1938) used this reasoning in calculating the distance-specific, incremental economy for the cost of human walking and running. In 1950, Gabrielli and Von Karman used a similar analysis to evaluate the efficiency of a variety of vehicles that differed in mass. Tucker (1970) and Schmidt-Nielsen (1972) used this efficiency index to compare runners, fliers and swimmers. Taylor *et al.* (1970) applied this calculation to locomoting mammals and found that distance-specific, incremental economy attains a minimum at high speeds (Fig. 7.2F). Taylor *et al.*, (1970) termed this index the minimum cost of transport (C_{min}). C_{min} represents the minimum metabolic cost required by an animal to travel a given distance. Mathematically, it equals the slope of the V_{O2ss} versus speed relationship (Fig. 7.2E).

Body mass and form affect C_{min}, but temperature appears to have little effect.

Body mass. Surprisingly, differences in body mass can account for much of the variation in C_{min} (Fig. 7.5A; Full, 1989; Herreid, 1981). When comparing over 150 species of animals that vary in leg number or position, body shape, and skeletal type, it is striking that larger animals of nearly all forms require less metabolic energy on a per gram basis to travel a given distance than do smaller animals. Birds and mammals (Taylor *et al.*, 1970; Fedak and Seeherman, 1979; Taylor *et al.*, 1982), lizards (John-Alder *et al.*, 1986), salamanders (Full *et al.*, 1988), crustaceans (Herreid and Full, 1988), insects (Herreid and Full, 1984; Herreid *et al.*, 1981; Jensen & Holm-Jensen, 1980; Lighton, 1985; Lighton *et al.*, 1987), and myriapods (Full and Herreid, 1986) follow a similar relationship. Moreover, studies on ghost crabs differing in body mass (2–70 g)

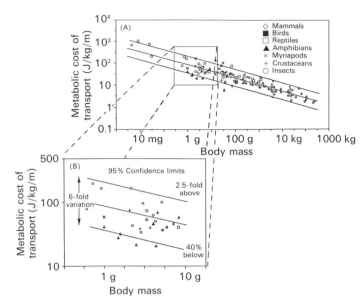

Fig. 7.5. Log mass-specific minimum cost of locomotion as a function of body mass. (A) Despite variation in body form the mass specific minimum cost of locomotion decreases with an increase in body mass over seven orders of magnitude in body mass. (B) Inset. Significant variation in the minimum cost of locomotion exists at an given body mass. Lines represent 95% confidence intervals. Data from various sources (see Full, 1989).

suggest that this same relationship may apply intraspecifically as well (Full, 1987).

Temperature. In contrast to gross and net economy, temperature appears to have little affect on incremental economy or C_{min}. For example, Madagascar hissing cockroaches, as well as two other cockroach species require the same minimum amount of metabolic energy to move a given distance at 15, 25 or 35°C (Herreid *et al.*, 1981). Temperature independence of C_{min} has also been demonstrated in several species of lizards (John-Alder and Bennett, 1981; Taylor, 1977). An explanation for this temperature independence is not yet available.

Body form - species differences. Considerable interspecific variation in C_{min} does exist at a given body mass (Full *et al.*, 1989). For example, two insects of the same body mass measured at the same temperature can differ in C_{min} by two–fold (Full *et al.*, 1990). Fedak and Seeherman (1979) have noted that cursorial mammals tend to have a low C_{min} relative to other mammals. Penguins clearly require more energy when waddling than other walking birds (Pinshow *et al.*, 1977). Red kangaroos require less energy to hop than comparably sized animals (Dawson and Taylor, 1973). Gila monsters have a relatively lower C_{min} than other lizards of the same body mass (John-Alder *et al.*, 1982). Variation in morphology and physiology can result in a six-fold variation in the minimum cost of locomotion at any given body mass (Fig. 7.5B). Costs 2.5–fold above and 40% below the predicted rate for a given body mass fall within the confidence limits.

7.5.5   E. Conclusions - economy

Comparable measures of economy can be obtained on diverse species. However, more effort must be made at quantifying the determinants of energy expenditure and less effort should be directed toward refining baselines which have no physiological bases. Much of the variation in economy can be explained by body mass and temperature. Striking general trends in economy are apparent over eight orders of magnitude in body mass. Yet, significant variation still exists among species when the effects of body mass and temperature are removed.

7.6   MECHANICAL ENERGY OUTPUT

Comparing a broader notion of efficiency (i.e. effectiveness) with the more limited definition of mechanical efficiency is one reason for the mismatch between what we hypothesize as efficient and what has been measured. It is not the only reason. A mismatch may still occur even when we are referring to mechanical efficiency. The major reason for the mismatch is related to the complexity involved in locomotion. It is this complexity which makes the determination of the link between mechanical energy output and metabolic energy input extremely difficult to identify. The behavior of the whole system cannot be predicted easily from the sum of the parts studied in isolation. Muscles, nerves, skeletal and support elements, all must act in concert to allow movement.

What can measures of mechanical energy output reveal about the consequences of variation in morphology and physiology for legged runners? Whole animal mechanical energy output will be most useful if: (1) we can obtain comparable measures in diverse species, (2) significant variation in mechanical energy output exists, and (3) differences in metabolic energy input are produced by variations in mechanical energy output.

7.6.1   *A. What is the mechanical energy output?*

An animal moving at a constant speed on the level can be said to do no mechanical work (i.e. positive increases and negative decreases in energy fluctuations cancel). A zero mechanical efficiency for steady-state, terrestrial locomotion is misleading. If animals moved like rolling wheels at a constant speed, then little mechanical energy would be necessary to move, since drag appears to be small in comparison to swimming and flying. However, during locomotion on land, both the body and limbs undergo repeated accelerations and decelerations during each cycle. Only the average speed of the whole animal is constant when determined over several strides. The mechanical energy (i.e. potential and kinetic) used to accelerate the body or center of mass is referred to as external energy (E_{ext}; Fig. 7.1; Fenn, 1930; Cavagna, 1975). Internal energy (i.e. E_{int}) represents the energy used to rock the trunk and swing the limbs relative to the body's motion.

7.6.2   *B. Increases in mechanical energy - counting energy more than once?*

(1) Energy transfer - Is what you see what you pay for?   Movements corresponding to external and internal energy changes are what we observe during locomotion. We usually assume that the energy generated to produce these movements is supplied by locomotor muscles during each step. Increases in mechanical energy originate from concentric contractions during which muscles shorten. Muscles which shorten in the direction of the applied force have accomplished positive mechanical work (Fig. 7.1). Increases in mechanical energy resulting from concentric contractions require metabolic energy. Unfortunately, summing the positive or absolute increases in kinetic, potential and rotational energy for each segment of the body and limbs will greatly overestimate the energy that must be supplied repeatedly by muscles and is aptly termed 'pseudowork' (Pierrynowski *et al.*, 1980; Williams, 1985). One source of the overestimate results from energy conservation and its transfer both within and between segments (Fig. 7.1). In humans one-third of the positive energy changes measured could be the result of exchanges within segments and one-third could result from exchanges between adjacent

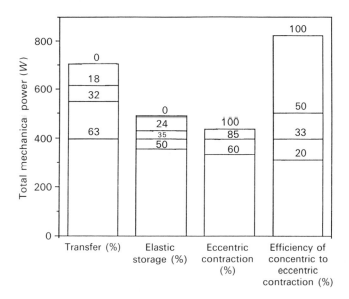

Fig. 7.6. The effect of various assumptions on the total mechanical power during running in humans. The percent energy transfer, elastic storage, eccentric contraction, and efficiency of concentric to eccentric contraction all significantly affect mechanical power output estimates. Total mechanical power = [(1 - %transfer/100)(1 - % elastic storage/100)* E_{pos}] + [% eccentric contraction * E_{neg}/% efficiency of contraction type] where E_{pos} equals the sum of the increase in mechanical energy and E_{neg} equals the sum of the decreases. Adopted from Williams and Cavanagh (1983).

segments (Pierrynowski *et al.*, 1980). The situation is analogous to determining the energy in a whip delineated into many segments. Energy, input from your hand, travels down the rope and causes each segment to move. If each segment had the capability of generating its own movement (i.e. had muscles that were active), then we might calculate the energy in the whip by summing the energy involved in moving each segment separately. Actually, each segment can be moved by the transfer of energy from the previous segment. Additional energy is not required for the movement of each segment. If energy transfer is complete, then the total energy can be determined from only one segment. Fig. 7.6 shows that mechanical power output estimates in humans can vary by 75% depending on the degree of transfer (Williams and Cavanagh, 1983).

During walking, many birds and mammals (Cavagna *et al.*, 1977; Heglund *et al.*, 1982a), as well as eight-legged crabs (Blickhan and Full,

1987), can transfer forward kinetic energy to gravitational potential and vice versa, much like an egg rolling end over end. As the body oscillates up and down, energy transfer or recovery can be as large as 50–70%. Summing the increases in forward kinetic and gravitational potential energy separately for a rolling egg would assume no transfer and lead to an obvious overestimate in the energy input for each cycle of rolling.

(2) Elastic strain energy   During running, energy can be stored temporarily as elastic strain energy (E_e) in tendons and other musculoskeletal structures and later transferred to potential and kinetic energy of the body (Fig. 7.1; Alexander, 1984). When elastic strain energy is stored, muscles contract as tendons are stretched. Elastic strain energy is released when the same muscles then do work and shorten. Muscles and tendons operating in this mode are analogous to a spring of a pogo stick or a bouncing ball. If elastic strain energy were ignored and mechanical energy determined by summing the increases in kinetic and gravitational potential energy for each step or stride, it could greatly overestimate the energy that must be generated by the muscles. In humans, mechanical power output can vary by as much as 40% depending on the amount of elastic storage (Williams and Cavanagh, 1983; Fig. 7.6).

Variation in body mass and form can potentially affect the amount of strain energy that can be stored.

Body mass. The contribution of elastic storage and recovery (E_e) as a function of body mass remains unclear. However, large vertebrates must be able to store and recover considerable amounts of energy during locomotion, since whole body efficiencies greatly exceed estimates of peak isolated muscle efficiency ($> 25\%$: Cavagna *et al.*, 1964, 1976, 1977; Heglund *et al.*, 1982b). The ability of small animals to store energy may be more limited. Kangaroo rats do not have the capacity to store as much elastic strain energy as kangaroos because they have relatively thicker tendons (Biewener *et al.*, 1981). It is not known whether small arthropods, such as cockroaches and crabs, are similarly limited in using their muscle, apodemes or other skeletal material as springs. Surprisingly, small cockroaches and eight-legged crabs have ground reaction force patterns and energy fluctuations that suggest the use of a bouncing or running gait comparable to that found in mammals (Fig. 7.7). Moreover, these animals attain a maximum sustainable stride frequency of the same magnitude and at the same speed predicted by scaling relationships of larger running mammals, despite the striking diversity

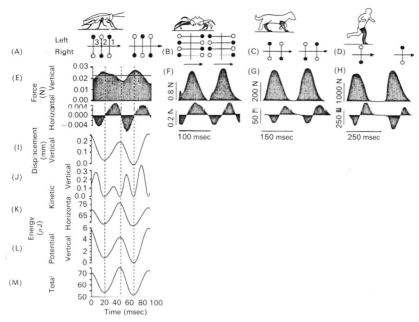

Fig. 7.7. Gait patterns of a cockroach (A), crab (B), dog (C) and human (D) during one stride. Filled circles represents a leg contacting the ground, whereas open circles represent legs moving in the air. Each animal can propel itself by two alternating sets of legs (i.e. 1–4 legs per set). Cockroaches use an alternating tripod gait. Left hind, right middle and left front alternate with the right hind, left middle and right front. Vertical and horizontal ground reaction forces for a running cockroach (E), ghost crab (F, Full, 1987), dog (G, Cavagna *et al.*, 1977) and human (H, Cavagna *et al.*, 1977). Segments represent one stride period. When vertical force equals zero the animal displays an aerial phase. Positive horizontal force represents braking, whereas negative values show acceleration of the center of mass forwards. Vertical displacement of the center of mass (I), vertical (J) and horizontal (K) kinetic energy, gravitational potential energy (L), and the sum of the three energies (M) for one stride of the cockroach.

in morphology and physiology (Full, 1989). Perhaps some storage can occur in these arthropods which operate their legs at high frequencies. Arthropods specialized for jumping, such as locusts and fleas, can store energy in apodemes and in resilin pads of joints (Bennett-Clark and Lucey, 1967; Bennett-Clark, 1975).

Body form - species differences. Since E_e is so difficult to quantify, it is not yet possible to know the extent of variation present for animals of similar size that vary in structure. In kangaroos metabolic energy actually decreases as speed increases (Dawson and Taylor, 1973).

As much as a 59% saving may result from elastic storage (Alexander, 1982; Alexander and Vernon, 1975; Cavagna *et al.*, 1977). Cavagna *et al.* (1964) estimate that running humans elastically conserve approximately 35% of the energy that must otherwise be supplied by muscles. Alexander (1984) suggested that in some galloping mammals the back might incorporate a spring that could store and return energy. Obviously, E_e requires further study, especially in small animals.

7.6.3   C. Decreases in mechanical energy

Surprisingly, decreases in mechanical energy can also require metabolic energy. Muscles that lengthen or are stretched while contracting are said to do negative mechanical work. Actually, muscles that undergo these eccentric contractions when lengthened are absorbing energy. Eccentric contractions require metabolic energy (Fig. 7.1). Since the metabolic costs of eccentric contractions are associated with decreases in energy, then perhaps mechanical energy output should be best estimated by summing the absolute value of positive and negative changes in mechanical energy. This approach has been used in the study of human locomotion (Pierrynowski *et al.*, 1980; Williams, 1985).

Summing the absolute value of the energy changes assumes that (1) all decreases in mechanical energy result in eccentric contractions and (2) the metabolic energy cost of negative and positive work is equal. Decreases in mechanical energy can also result from joint range limitations and muscle viscosity (Fig. 7.1). Mechanical power output in humans can vary by 20% if nearly all versus two-thirds of the decreases in mechanical power are associated with eccentric contractions (Fig. 7.6). In addition, several studies have shown that the metabolic costs of concentric contractions differ from eccentric contractions. Running uphill requires more positive work by muscles contracting concentrically, and more metabolic energy, than running downhill which demands more negative work by eccentrically contracting muscles. Eccentric contractions require one-third to as little as one-fifth of the metabolic energy of concentric contractions (Margaria, 1968; Williams and Cavanagh, 1983). Since the metabolic energy costs of negative and positive work are not equal, then perhaps mechanical energy output should be best estimated by weighting the negative changes in energy to reflect the lower cost. Williams and Cavanagh (1983) found a six-fold variation in the mechanical power output estimate depending on the relative cost selected for positive versus negative work (Fig. 7.6).

7.6.4   D. External energy - mechanical energy changes of the center of mass

External mechanical energy changes of the body or center of mass represent a major portion of the mechanical energy output of animals (Cavagna *et al.*, 1977; Heglund *et al.*, 1982a). Comparable estimates have been obtained in diverse species from measurements of the ground reaction forces. Surprisingly, vertical and horizontal ground reaction force patterns can be similar in two-, four-, six- and eight-legged runners (Fig. 7.7). Mechanical energy changes of the center of mass are derived from integration of the ground reaction forces. Most determinations of total mechanical energy change of the center of mass represent the instantaneous sum of the kinetic and potential energy fluctuations assuming complete transfer among energies. The external power of the center of mass (E_{ext}) has been calculated from the sum of the positive increases in the total energy assuming no eccentric contractions or ones of very low cost. Elastic storage has been assumed to be zero. In birds, mammals, crabs and cockroaches, E_{ext} increases linearly with speed and extrapolates to near zero at zero speed (Blickhan and Full, 1987; Cavagna *et al.*, 1977; Full and Tu, 1989; Heglund *et al.*, 1982a). Therefore, distance-specific E_{ext} is independent of speed and is analogous to distance-specific, incremental metabolic cost or the minimum cost of transport (Fig. 7.2F).

Body mass. Differences in body mass can account for much of the variation in E_{ext}. Despite variation in body shape and skeletal type, E_{ext} is directly proportional to body mass over a wide range of speeds (Blickhan and Full, 1987; Cavagna *et al.*, 1977; Full and Tu, 1989; Heglund *et al.*, 1982a). Studies on ghost crabs suggest that a similar trend may be present for animals of the same species that differ in mass (Blickhan and Full, 1987). In contrast to mass-specific, $V_{O_2 gross}$ (Fig. 7.4), mass-specific E_{ext} follows a similar function of speed for animals that differ in body mass (Fig. 7.8). The E_{ext} generated to move an animal one meter is nearly directly proportional to body mass or independent of body mass when represented on a mass-specific basis.

Body form - species differences. The mechanical power used to accelerate an animal's center of mass upwards and forwards (E_{ext}) has only been measured in about 15 species (Blickhan and Full, 1987; Cavagna *et al.*, 1977; Full and Tu, 1989; Heglund *et al.*, 1982a). The energy generated to move a kilogram of mass one meter is approximately 1.1 J/kg/m and varies by 50% (i.e. upper and lower 95% confidence limits).

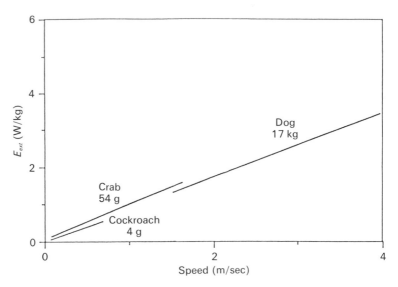

Fig. 7.8. Mass-specific mechanical energy of the center of mass (external energy) as a function of speed for cockroaches (4 g; Full and Tu, 1989); crabs (30–90 g; Blickhan and Full, 1987); and dogs (17 kg; Heglund *et al.*, 1982a).

This variation is much smaller than that measured for metabolic power. Surprisingly, animals with many legs due not appear to have smooth rides during which little acceleration and deceleration occur (Fig. 7.9; Blickhan and Full, 1987; Full and Tu, 1989).

7.6.5   E. Internal energy - limb and trunk motion relative to the center of mass

The internal energy necessary to accelerate the limbs and the body relative to the center of mass (E_{int}) increases curvilinearly with speed in birds and mammals (Fedak *et al.*, 1982). In most species, E_{int} becomes a greater portion of total mechanical energy as speed increases. Values will vary depending whether no transfer (0%), transfer between only adjacent segments (18%), transfer within segments (not between limbs; 32%), or maximum transfer (63%) is assumed (estimates on humans from Williams and Cavanagh, 1983; Fig. 7.6).

Body mass. Internal energy increases as a function of body mass in birds and mammals (Fedak *et al.*, 1982). Swinging larger limbs and

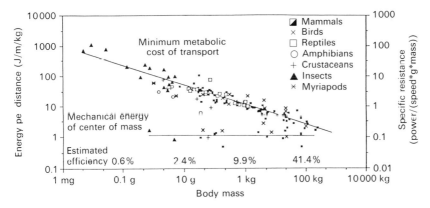

Fig. 7.9. Logarithmic plot of mass-specific external mechanical energy (E_{ext}) and minimum metabolic energy (C_{min}) used to move one kg of animal one meter in distance. Small animals, regardless of form, require relatively more metabolic energy to travel a distance than large animals, but do not produce relatively more mechanical energy to move their center of mass. Data for mammals and birds, lizards, amphibians, crustaceans, insects and myriapods are from various sources (see Full, 1989, for references). $C_{min} = 10.8\ M^{-0.32}$ (r2 = 0.87) and $E_{ext} = 1.07\ M^{-0.01}$.

trunks is associated with a greater E_{int} than swinging smaller limbs and rocking smaller bodies.

Body form - species differences. Variation in E_{int} among the few birds and mammals studied is significant. E_{int} ranged from 15% of the estimated total mechanical energy in a quail to 54% in a dog at the fastest speeds tested (Fedak *et al.*, 1982; Heglund *et al.*, 1982b). Mass-specific E_{int} does not show any regular function related to leg number or structure. Theory used to design walking machines predicts that an increase in leg number may actually reduce E_{int} (Hirose and Umetani, 1978). It also predicts that the 'knee above the hip' posture of most arthropods may decrease E_{int} compared to a four-legged animal with an upright stance (Kaneko *et al.*, 1987). Many-legged animals, such as crabs, undoubtedly generate considerable internal energy to accelerate their limbs and body relative to their center of mass, especially at fast speeds. Total mechanical energy would be better estimated in a 50 - g crab by increasing E_{ext} 7% at the lowest speeds and as much as 32% at maximum speed, just as it would for a mammal of the same body mass (Blickhan and Full, 1987; Cavagna *et al.*, 1977; Full and Tu, 1989; Heglund *et al.*, 1982a). Further and more rigorous estimates of E_{int}

are obviously necessary in many-legged animals, such as insects and myriapods.

7.6.6   F. Conclusions - mechanical energy output

Mechanical energy output will vary depending on: whether external and internal work are both determined; what assumptions are made concerning energy transfer within and between segments; the amount of elastic strain energy stored; the degree of eccentric contractions; and the relative efficiency of positive and negative work. Despite the potential for variation, the amount of energy per unit mass generated to move the center of mass one meter (E_{ext}) is remarkably similar for species that vary in size and form.

7.7   INSIGHTS FROM EFFICIENCY MEASURES - ALL EFFICIENCIES ARE NOT CREATED EQUAL

The mechanical efficiency of terrestrial locomotion can be calculated using one of at least four values for mechanical energy as the numerator and one of at least three values for metabolic energy input as the denominator. The number of different definitions for efficiency begins to rival those found in ecology. They include gross, net, incremental, instantaneous, work, apparent, muscle, muscular, pseudo, and center of mass efficiencies (Alexander, 1977; Donovan and Brooks, 1977; Gaesser and Brooks, 1975; Goldspink, 1977; Stainsby *et al.*, 1980; Taylor, 1980; Winter, 1979). The value selected for the numerator, mechanical energy output, will vary depending on: (1) whether external and internal work are both determined, (2) what assumptions are made concerning energy transfer within and between segments, (3) the amount of elastic strain energy stored, (4) the extent of eccentric contractions, and (5) efficiency of positive versus negative work. The value selected for the denominator, metabolic energy input, will vary depending on the assumptions made concerning the baseline energy during exercise (i.e. gross, net, incremental economy).

All efficiencies values represent whole body efficiency and are global measures. Insufficient evidence is available to claim that any ratio represents actual isolated muscle efficiency, although some values are undoubtedly better estimates than others. The inequality of whole body versus muscle efficiency has been pointed out several times in the past. In fact, this recognition has lead to terminology for whole body efficiency, such as muscular efficiency, as opposed to muscle efficiency (Stainsby *et al.*, 1980). This being the case, then what can whole body measures of efficiency tell us? Can these whole body measures reveal the consequences of variation in morphology and physiology that affect locomotion on land?

7.7.1 A. Does variation in mechanical energy output produce concomitant variation in metabolic energy input?

Do animals that have a lower mechanical power output consume less oxygen during exercise than those with a higher mechanical power output? If so, how much less? Muscles would be expected to use less metabolic energy if work was decreased by: effective storage of elastic strain energy and energy transfer among segments; reduced accelerations and decelerations of the center of mass in the horizontal, vertical and lateral directions; reduced limb moments of inertia produced by a decrease in limb mass or concentration of mass near the center of rotation.

Body mass. In birds and mammals variation in economy related to body mass cannot be completely explained by concomitant variation in mechanical energy output (Heglund *et al.*, 1982b). Larger animals use less metabolic energy to move a gram of body mass one meter than do small animals. Over orders of magnitude of five to seven in body mass, mass-specific metabolic cost (C_{min}) varies by more than two to three orders of magnitude, whereas the mass-specific, mechanical power (E_{ext}) generated to move a gram of animal one meter is relatively independent of body mass (Fig. 7.9). Therefore, whole body efficiency increases with body size. Large birds and mammals have efficiencies that exceed 30–50%. At least part of the reason for whole body efficiencies exceeding peak isolated muscle efficiency (i.e. 25–30%) is the inability to account for elastic storage and transfer in the mechanical energy estimate. Energy that is stored and transferred is instead calculated to be generated by muscles. If large animals store much more elastic strain energy, this may explain some, but not all, of the variation in whole body efficiency with body mass.

The different scaling of metabolic and mechanical power cannot be completely explained by any reassesment of the mechanical power estimate. Large animals (100 kg) exhibit approximately three to five-fold differences between metabolic and mechanical power (Fig. 7.9). Our recent estimates of mechanical power output show that this difference reaches 100–fold in small insects (0.07–4 g; Full and Tu, 1989, submitted; Fig.7.9). Maximum mechanical power output estimates for small animals (1 g; 0% transfer, 0% elastic storage; 85% eccentric contraction and 100% or equal efficiency of con- and eccentric contractions) would have to be increased by 20–30-fold to attain a whole body efficiency of 25%. Maximum mechanical power output estimates for small animals combined with minimum mechanical power output estimates for large animals (100 kg; 63% transfer; 60% elastic storage; 37% eccentric contractions and 20% efficiency of con- and eccentric contractions) only increases the difference in mechanical power between large and small animals to seven–fold, not nearly the 50–fold necessary to make efficiency independent of animal size (i.e. parallel relationship between the functions relating the minimum metabolic cost of transport and mechanical energy output to body mass; Fig 7.9).

Body form - species differences. Fewer efficiency data are available for animals of similar body mass. In kangaroos the low metabolic cost appears to result from a reduced mechanical power output (Dawson and Taylor, 1973). Kangaroos store significant amounts of elastic strain energy (Alexander and Vernon, 1975; Cavagna *et al.*, 1977). The elevated metabolic cost of waddling penguins and geese is most likely correlated with an increase in mechanical power, but actual output measurements are lacking (Pinshow *et al.*, 1977; Baudinettee and Gill, 1985). Williams and Cavanagh (1985) found that humans with a lower $V_{O2gross}$ were more effective at deriving power from energy transfer. Increased mechanical work due to loading the limbs is directly proportional to the increase in metabolic cost in humans (Martin, 1985; Myers and Steudel, 1985). When segments are loaded more distally, oxygen consumption increases as a function of the increase in mechanical power output.

More evidence is available that suggests a weak correlation between variation in economy or metabolic energy input and mechanical energy output. Cheetahs, gazelles and goats differ considerably in limb configuration, and presumably in the mechanical energy necessary to swing their limbs. Yet, Taylor *et al.* (1974) found no significant difference in oxygen consumption, at least at low speeds. Likewise, studies of animals running with loaded packs show a proportional increase in $V_{O2gross}$ with the load carried, a much greater metabolic cost than predicted if the metabolic cost of swinging the limbs was a very large fraction of the total cost (Taylor *et al.*, 1980). In humans, significant variation in economy of elite versus good distance runners does not correlate with mechanical parameters, except stride frequency (Cavanagh *et al.*, 1977; Pollack, 1977). In our laboratory we varied morphology and presumably mechanical energy output directly by producing quadrupedal cockroaches from six-legged animals (i.e. removed middle legs). Four-legged cockroaches wobble or roll considerably during locomotion. Surprisingly, four-legged runners do not show any significant difference in metabolic cost at high speeds when compared to six-legged runners (Full and Pham, unpublished).

If variation in metabolic energy input is produced primarily by variation in mechanical energy output, then animals with relatively low mechanical energy output should require less metabolic energy when the effects of body mass are removed for both energy values. My analysis of E_{ext} and C_{min} residuals does not show any correlation when the effects of body mass are removed. Animals with many legs or seemingly awkward running styles, such as crabs and cockroaches, do not show variation in mechanical power that is accompanied by a corresponding change in metabolic cost (Full, unpublished results). This may be in large part due to the inadequate sample size. Moreover, estimates of elastic strain energy and transfer are needed to determine if increases in the mechanical power used to accelerate an animal's center of mass upwards and forwards correlate with an elevated C_{min}.

7.7.2   B. What can explain the variation in economy or metabolic energy input in addition to mechanical energy output?

Metabolic cost can vary even when little difference in mechanical energy output is apparent. Why does such variation in whole animal efficiency exist? In 1980, Taylor suggested that the cost of muscle force production might determine the metabolic cost of locomotion. Many locomotor muscles function primarily as force generators and undergo near isometric contractions (i.e. average zero shortening velocity), especially when stabilizing joints and maintaining a running posture while supporting the body's weight. Moreover, muscles are active when they are stretched and absorb energy. Muscles functioning in these ways require metabolic energy without production of positive mechanical work (Fig. 7.1). The hypothesis that the cost of muscle force production determines locomotor cost is consistent with previous research on isolated muscle which has shown a good correlation between metabolic cost and the area under the muscle force versus time curve, the time-tension integral (Stainsby and Fales, 1973). Whole animal metabolic cost may best be explained by: (1) the rate of force production (Taylor, 1985); and (2) the total amount of force produced.

(1) Rate of force production **Body mass**. Taylor *et al.* (1980) used the 10–fold variation in C_{min} with body mass found in mammals to test the hypothesis that the metabolic cost of locomotion is determined by the cost of force production. They exercised animals loaded with back-packs on a treadmill. No change in acceleration of the center of mass was observed between loaded and unloaded animals. Therefore, muscle force increased in direct proportion to the load added. Oxygen consumption also rose in direct proportion to the added load for animals which ranged in mass from a rat to a horse. Since small animals have higher mass-specific metabolic costs for unloaded running (see Fig. 7.3), an equivalent increase in load or force produced a much greater increase in mass-specific metabolic cost in small animals compared to large. The development of each Newton of force by a small animal appears to require more metabolic energy than the development of the same amount of force by a large animal.

Small animals seem to require more metabolic energy to move a gram of body mass than larger ones because they must turn their muscles on and off more frequently per unit time or distance. The metabolic cost of force production varies with body mass in a similar manner to stride frequency (Fig. 7.10). Higher rates of contraction appear to result in additional cost due to more frequent activation (due to Ca^{++} movement; Rall, 1986) and the higher costs associated with the more rapid cycling of cross-bridges (Heglund and Cavagna, 1987). When the metabolic cost of locomotion is normalized for the rate of force production, the mass-specific metabolic cost of locomotion for one stride is remarkably independent of body mass (Heglund and Taylor, 1988). Therefore, the greater metabolic cost per unit mass of small animals to travel a given distance could be explained by the fact that small animals, with shorter legs than larger animals, must take more steps costing an equivalent amount of mass-specific metabolic energy to cover the same distance.

Body form - species differences. Full *et al.* (1990) found significant variation in C_{min} of one gram insects that differ in form (i.e. leg configuration). Caterpillar hunting beetles use only half the energy of field crickets and American cockroaches of the same mass. Normalizing for the rate of muscle force production by determining the metabolic cost per stride fails to account for the interspecific variation in the cost of locomotion in insects of the same mass as it does for mammals that differ in mass (Fig. 7.11). Kram and Taylor (1989) suggested that normalizing for the rate of force production by using ground contact time may be more appropriate than dividing by stride frequency, because

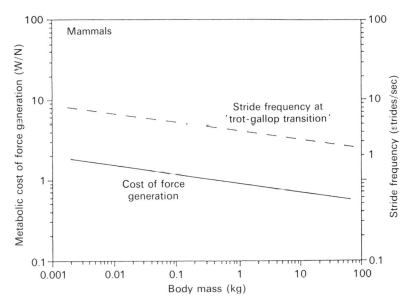

Fig. 7.10. Log body mass versus log stride frequency and the metabolic cost of force generation at equivalent speeds (e.g. trot-gallop transition for mammals; Heglund *et al.*, 1974). The metabolic cost of generating a Newton of force in a small animal is greater than in large animals and parallels the metabolic cost of locomotion (Fig. 7.9). The cost of force generation may be greater in small animals because of added costs due to more frequent activation (due to Ca^{++} movement) and the higher costs associated with the more rapid cycling of cross-bridges. Consistent with the idea that higher costs are associated with higher cycling rates are the data on stride frequency that parallel the cost of force generation (Taylor, 1985).

muscles are developing most of their force when the legs are in contact with the ground and are supporting the body's weight. Although the ground contact cost in insects (1.5–3.1 J kg^{-1}) was similar to that measured in mammals (2 J kg^{-1}), it also does not explain the interspecific variation in the cost of locomotion (Fig. 7.11).

(2) Amount of force production **Body size**. Variation in the total amount of muscular force generated to produce the same ground reaction force could lead to variation in metabolic cost. Biewener (1983, 1989) has found a 10–fold decrease in the effective mechanical advantage of mammalian limbs (i.e. due to variation in lever arms) with a decrease in body mass. Small mammals have a more crouched posture during locomotion than larger mammals which requires relatively greater mus-

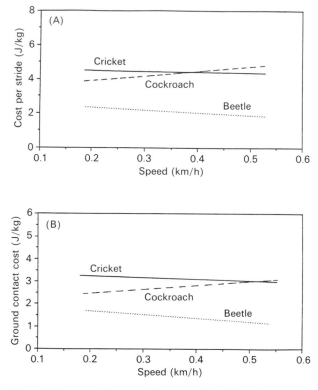

Fig. 7.11. Metabolic cost of locomotion normalized for the rate of force production in crickets, cockroaches and beetles of similar mass (1 g). (A). The metabolic cost of locomotion per stride was independent of speed, but was significantly lower in beetles than in crickets or cockroaches. (B). Ground contact cost was independent of speed and was significantly lower in beetles than in crickets or cockroaches. Ground contact cost was calculated by multiplying steady-state oxygen consumption by contact time (Full *et al.*, 1990). 1 ml O_2 = 20.1 J.

cle force production. Small mammals generate the additional force with a relatively greater cross sectional area of active muscle, keeping muscle stress relatively constant (Perry *et al.*, 1988). However, the larger force per volume of activated muscle in small mammals should result in metabolic costs which are relatively greater than in large mammals. Variation in mechanical advantage could also explain the variation in the metabolic cost of locomotion with body mass in mammals.

Body form - species differences. Variation in mechanical advantage could very well explain variation in the cost of locomotion in different species of the same body mass. Full *et al.* (1990) could not account for differences in metabolic cost of three similarly sized insects that differed in limb configuration by normalizing for variation in the rate or cost of force production (Fig. 7.11). Effective mechanical advantage and the amount of muscle force required to produce the same ground reaction force may explain the variation in the metabolic cost. The low cost of locomotion in the beetle may result from a greater effective mechanical advantage of leg muscles in the limbs compared to the cockroach and cricket. Preliminary estimates of locomotor muscle mass support this contention.

7.8 CONCLUSIONS

Efficiency, as applied to land locomotion, is used in two ways (1) mechanical efficiency, and (2) competency of performance. Unfortunately, both uses of the term represent many different numerical indices. One solution to this problem would be to stop using the general term 'efficiency' because of its ambiguity. This strategy is logical, but impractical. A better approach would be to state clearly its operational definition when using the term and to be skeptical when claims of efficiency are made without a precise definition.

(1) Whole animal efficiency used as a measure of effective performance is useful and can aid in our search for the mechanistic bases of how diverse animals move on land, as well as provide data for ecological and evolutionary studies. Steady-state, constant speed locomotion is only one of the 'events' that requires further study. Others include: intermittent activity, obstacle negotiation (e.g. climbing, leaping and maneuvering), and burst locomotion. Economy is only one measure of effective performance. Other performance measures which are receiving increasing attention include: endurance or fatigue resistance, distance travelled, acceleration and speed, durability and strength, and maneuverability and stability.

(2) The economy of locomotion is one performance measure that has been determined frequently. Variation in economy exists in resting, offset and incremental components. Functional explanations of these components are difficult because of our lack of data on the amount of energy required by non-locomotor sources during exercise. Less effort should be directed to refining baselines and more effort should be made at quantifying the determinants of energy expenditure. Much of the variation in

economy can be accounted for by variation in body mass and temperature. Yet, variation in body form is equally important. The minimum cost of locomotion at a given body mass or species differences can vary by as much as six–fold. This variation remains to be explained.

(3) Mechanical energy output provides only one source of variation in the economy of locomotion. It is only one link between performance and the structure and function of muscles and skeletal structures. Mechanical energy output will vary depending on: whether external and internal work are both determined, what assumptions are made concerning energy transfer within and between segments, the amount of elastic strain energy stored, the degree of eccentric contractions, and the relative efficiency of positive and negative work.

(4) Whole animal mechanical efficiency appears to be highly variable among animals that differ in body form and size and not simply equal to the efficiency of isolated muscle. Investigators comparing the mechanical efficiency among animals who propose hypotheses for reported differences must carefully consider the sources of variation in both mechanical energy output and metabolic energy input. The movements we observe are not necessarily paid for by the animal based on the assumption of a constant whole body mechanical efficiency. Mechanical energy can be stored and released, not requiring additional metabolic cost, or muscles may contract nearly isometrically and demand metabolic energy without producing detectable mechanical work.

(5) Variation in the economy of locomotion may be best explained by the differences in the cost of muscle force production and the total force produced.

7.9   REFERENCES

Alexander, R. McNeill. (1977). Terrestrial locomotion. In: *Mechanics and Energetics of Animal Locomotion* (R. McNeil Alexander & T. Goldspink, eds.), Wiley & Sons, New York, pp. 168–203.

Alexander, R. McNeill. (1982). *Locomotion of Animals*. Blackie, Glasgow.

Alexander, R. McNeill. (1984). Elastic energy stores in running vertebrates. *Amer. Zool.* **24**, 85–94.

Alexander, R. McN. & Vernon, A. (1975). Mechanics of hopping by kangaroos (Macropodidae). *J. Zool., Lond.* **177**, 265–303.

Anderson, J.F. & Prestwich, K.N. (1985). The physiology of exercise at and above maximal aerobic capacity in a theraphosid (tarantula) spider, *Brachypelma simihi* (F.O. Pickard-Cambridge). *J. comp. Physiol.* **155**, 529–39.

Arnold, S. J. (1983). Morphology, performance and fitness. *Amer. Zool.* **23**, 347–61.

Baudinette, R.V. & Gill, P. (1985). The energetics of 'flying' and 'paddling' in water: locomotion in penguins and ducks. *J comp. Physiol.* **153**, 373–80.

Bennett, A.F. (1982). The energetics of reptilian activity. In: *Biology of Reptilia* (C. Gans & F.H. Pough, eds.), vol. 13., pp. 155–99, New York: Academic Press.

Bennett-Clark, H.C. & Lucey, E.C.A. (1967). The jump of the flea; a study of the energetics and a model of the mechanism, *J. exp. Biol.* **47**, 59–76.

Bennett-Clark, H.C. (1975). The energetics of the jump of the locust, *Schistocera gregaria*. *J. exp. Biol.* **63**, 53–83.

Biewener, A.A. (1983). Allometry of quadrupedal locomotion: the scaling of duty factor, bone curvature and limb orientation to body size. *J. exp. Biol.* **105**, 147–71.

Biewener, A.A. (1989). Design of the mammalian terrestrial locomotor system in relation to body size. *Bioscience* **39**, 766–83.

Biewener, A., Alexander, R. McNeill & Heglund, N.C. (1981). Elastic energy storage in the hopping of kangaroo rats (*Dypodomus spectabilis*). *J. Zool., Lond.* **195**, 369–83.

Biewener, A. A. & Blickhan, R. (1988). Kangaroo rat locomotion: design for elastic storage or acceleration? *J. exp. Biol.* **140**, 243–55.

Blickhan, R. & Full, R.J. (1987). Locomotion energetics of the ghost crab: II. Mechanics of the center of mass during walking and running. *J. exp. Biol.* **130**, 155–74.

Brooks, G.A., Donovan, C.M. & White, T.P. (1984). Estimation of anaerobic energy production and efficiency in rats during exercise. *J. Appl. Physiol.* **56**, 520–5.

Cain, S.M. (1971). Exercise O_2 debt of dogs at ground level and at altitude with and without -block. *J. Appl. Physiol.* **30**, 838–43.

Cavagna, G.A. (1975). Force platforms as ergometers *J Appl. Physiol.* **39**, 174–9.

Cavagna, G.A., Heglund, N.C. & Taylor, C.R. (1977). Mechanical work in terrestrial locomotion: two basic mechanisms for minimizing energy expenditure. *Am. J. Physiol.* **233**(5), R243-R261.

Cavagna, G.A., Saibene, F.P. & Margaria, R. (1964). Mechanical work in running. *J. Appl. Physiol.* **19**, 249–56.

Cavagna, G.A., Thys, H. & Zamboni, A. (1976). The sources of external work in level walking and running. *J. Physiol., Lond.* **262**, 639–57.

Cavagna, G.A. & Kram, R. (1985). Mechanical and muscular factors affecting the efficiency of human movement. *Med. Sci. Sports Exerc.* **17**, 326–31.

Cavanagh, P.R., Pollock, M.L. & Landa, J. (1977). A biomechanical comparison of elite and good distance runners. The marathon: physiological, medical, epidemiological and psychological studies. *Ann. N.Y. Acad. Sci.* **301**, 328–45.

Dawson, T. & Taylor, C.R. (1973). Energy cost of locomotion in kangaroos. *Nature, Lond.* **246**, 313–14.

Donovan, C.M. & Brooks, G.A. (1977). Muscular efficiency during steady-rate exercise. II. Effects of walking speed and work rate. *J. Appl. Physiol.* **43**,

431–9.

Fedak, M.A., Heglund, N.C. & Taylor, C.R. (1982). Energetics and mechanics of terrestrial locomotion: II. Kinetic energy changes of the limbs and body as a function of speed and body size in birds and mammals. *J. exp. Biol.* **79**, 23–40.

Fedak, M.A. & Seeherman, H.J. (1979). Reappraisal of energetics of locomotion shows identical cost in bipeds and quadrupeds including ostrich and horse. *Nature* **282**, 713–16.

Fenn, W.O. (1930). Work against gravity and work due to velocity changes in running. *Am. J. Physiol.* **93**, 433–62.

Full, R.J. (1986). Locomotion without lungs: energetics and performance of a lungless salamander, *Plethodon jordani. Am. J. Physiol.* **251**, R775–R780.

Full, R.J. (1987). Locomotion energetics of the ghost crab : I. Metabolic cost and endurance. *J. Exp. Biol.* **130**, 137–54.

Full, R.J. (1989). Mechanics and energetics of terrestrial locomotion: From bipeds to polypeds. In: *Energy Transformation in Cells and Animals* (eds. W. Wieser & E. Gnaiger). Thieme, Stuttgart, pp. 175–82.

Full, R.J. & Herreid, C.F. (1983). The aerobic response to exercise of the fastest land crab. *Am. J. Physiol.* **244**, R530-R536.

Full, R.J. & Herreid, C.F. (1984). Fiddler crab exercise: the energetic cost of running sideways. *J. Exp. Biol.* **109**, 141–61.

Full, R.J. & Herreid, C.F. (1986). Energetics of multilegged locomotion. *Proc. Int. Union Physiol. Sci.* **16**, 403.

Full, R.J., Herreid, C.F. & Assad, J.A. (1985). Energetics of the exercising wharf crab *Sesarma cinereum. Physiol. Zool.* **58**, 605–15.

Full, R.J., Anderson, B.D., Finnerty, C.M. & Feder, M.E. (1988). Exercising with and without lungs: I. The effects of metabolic cost, maximal oxygen transport and body and on terrestrial locomotion in salamander species. *J. Exp. Biol.* **138**, 471–85.

Full, R. J. & Tu, M.S. (1990). Mechanics of six-legged runners. *J. exp. Biol.* **148**, 129–46.

Full, R. J. & Tu, M.S. Mechanics of a rapid running insect: Two, four- and 6-legged locomotion. *J. Exp. Biol.* In press.

Full, R.J., Zuccarello, D.A. & Tullis, A. (1990). Effect of variation in form on the cost of terrestrial locomotion. *J. Exp. Biol.* **150**, 223-46.

Gabrielli, G. & von Karman, T. (1950). What price speed? *Mech. Eng.* **72**, 775–81.

Gaesser, G.A. & Brooks, G.A. (1975). Muscular efficiency during steady-rate exercise: effects of speed and work rate. *J. Appl. Physiol.* **38**, 1132–9.

Goldspink, G. (1977). Muscle energetics and animal locomotion. In: *Mechanics and Energetics of Animal Locomotion* (R. McN. Alexander & G. Goldspink, eds.), Chapman and Hall, London, p. 78.

Heglund, N.C. & Cavagna, G.A. (1987). Mechanical work, oxygen consumption and efficiency in isolated frog and rat striated muscle. *Am. J. Physiol.* **253**, C22-C29.

Heglund, N.C., Cavagna, G.A. & Taylor, C.R. (1982a). Energetics and mechanics of terrestrial locomotion. III. Energy changes of the center of mass as a function of speed and body size in birds and mammals. *J. exp. Biol.* **79**, 41–56.

Heglund, N.C., Fedak, M.A., Taylor, C.R. & Cavagna, G.A. (1982b). Energetics and mechanics of terrestrial locomotion. IV. Total mechanical energy changes as a function of speed and body size in birds and mammals. *J. exp. Biol.* **97**, 57–66.

Heglund, N.C. & Taylor, C.R. (1988). Speed, stride frequency and energy cost per stride: how do they change with body size and gait? *J. exp. Biol.* **138**, 301–18.

Heglund, N.C., Taylor, C.R. & McMahon, T.A. (1974). Scaling stride frequency and gait to animal size: mice to horses. *Science* **186**, 1112–13.

Hemmingsen, A.M. (1960). Energy metabolism as related to body size and respiratory surface, and its evolution. *Rep. Steno. Hosp.* **9**, 1 110.

Herreid, C.F. (1981). Energetics of pedestrian arthropods. In: *Locomotion and Energetics in Arthropods* (eds. C.F. Herreid & C.R. Fourtner), New York: Plenum Press pp. 491–526.

Herreid, C.F. & Full, R.J. (1984). Cockroaches on a treadmill: aerobic running. *J. Insect Physiol.* **30**, 395–403.

Herreid, C.F. & Full, R.J. (1985). Energetics of hermit crabs during locomotion: the cost of carrying a shell. *J. exp. Biol.* **120**, 297–308.

Herreid, C.F., Full, R.J. & Prawel, D.A. (1981). Energetics of cockroach locomotion. *J. exp. Biol.* **94**, 189–202.

Herreid, C.F. & Full, R.J. (1988). Energetics and locomotion. In: *Biology of the Land Crab* (eds. W. Burggren & B.R. McMahon), pp. 333–77. Cambridge University Press, New York.

Herreid, II, C. F., O'Mahoney, P.J. & Full, R.J. (1983). Locomotion in land crabs: respiratory and cardiac response of *Gecarcinus lateralis*. *Comp. Biochem. Physiol.* **74A**, 117–24.

Hirose, S. & Umetani, Y. (1978). *3rd CISM -IFToMM Symposium on Theory and Practice of Robots and Manipulators*, pp. 357–75.

Hoyt, D.F. & Kenagy, G.J. (1988). Energy costs of walking and running gaits and their aerobic limits in golden-mantled ground squirrels. *Physiol. Zool.* **61**, 34–40.

Hoyt, D.F. & Taylor, C.R. (1981). Gait and the energetics of locomotion in horses. *Nature, Lond.* **292**, 239–240.

Huey, R.B. (1987). Phylogeny, history and the comparative method. In: *New Directions in Ecological Physiology* (M.E. Feder, A.F. Bennett, W.W. Burggren & R.B. Huey, eds.), Cambridge University Press, Cambridge, pp. 76–97.

Huey, R.B. & Bennett, A.F. (1986). A comparative approach to field and laboratory studies in evolutionary biology. In: *Predator-prey Relationships* (M.E. Feder & G.V. Lauder, eds.), The University of Chicago Press, Chicago, pp. 82–98.

Hurst, R.J., Leonard, M.L., Watts, P.D., Beckerton, P. & Oritsland, O. (1982). Polar bear locomotion: body temperature and energetic cost. *Can. J. Zool.* **60**, 40–4.

Jensen, T.F. & Holm-Jensen, I. (1980). Energetic cost of running in workers of three ant species *Formica fusca* L., *Formica rufa* L., and *Camponotus herculaneus*

L. (Hymenoptera, Formicidae). *J. comp. Physiol.* **137**, 151–6.

John-Alder, H.B., & Bennett, A.F. (1981). Thermal dependence of endurance and locomotory energetics in a lizard. *Am. J. Physiol.* **241**, R342-R349.

John-Alder, H.B., Garland, T. & Bennett, A.F. (1986). Locomotory capacities, oxygen consumption, and the cost of locomotion of the shingle-back lizard (*Trachydosaurus rugosus*). *Physiol. Zool.* **59**(5), 523–31.

John-Alder, H.B., Lowe, C.H. & Bennett, A.F. (1982). Thermal dependence of locomotory energetics and aerobic capacity of the Gila Monster (*Heloderma suspectum*). *J. comp. Physiol.* **151**, 119–26.

Kaneko, M., Tachi, S., Tanie, K. & Abe, M. (1987). Basic study on similarity in walking machine from a point of energetic efficiency. *J. of Rob. and Auto.* **RA-3**, 19–30.

Lighton, J.R.B. (1985). Minimum cost of transport and ventilatory patterns in three African beetles. *Physiol. Zool.* **58**, 390–9.

Lighton, J.R.B., Bartholomew, G.A. & Feener, D.H. (1987). Energetics of locomotion and load carriage in the leaf-cutting ant *Atta colombica* Guer. *Physiol. Zool.* **60**, 524–737.

Margaria, R. (1938). Sulla fisiologia e specialmente sul consumo energetico della marcia e della corsa a varie velocita ed inclinazonioni del terreno. *Atti. Accad. Lincei Memorie, serie VI,* **7**, 299–368.

Margaria, R. (1968). Positive and negative work performance and their efficiencies in human locomotion. *Int. Ziet. feuer Angewandte Physiol. Einschliesslicj Arbietphysiologie.* **25**, 399.

Margaria, R., Cerretelli, P., Aghemo, P. & Sassi, G. (1963). Energy cost of running. *J. Appl. Physiol.* **18**, 367–70.

Martin, P.E. (1985). Mechanical and physiological responses to lower extremity

Meyers, M.J. & Steudel, K. (1985). Effect of limb mass and its distribution on the energetic cost of running. *J. exp. Biol.* **116**, 363–73.

Palidino, F.V & King, J.R. (1979). Energetic cost of terrestrial locomotion: biped and quadruped runners compared. *Can. Rev. Biol.* **38**, 321–3.

Perry, A.K. Blickhan, R., Biewener, A.A., Heglund, N.C. & Taylor, C.R. (1988). Preferred speeds in terrestrial vertebrates: are they equivalent? *J. exp. Biol.* **137**, 207–19.

Pierrynowski, M.R., Winter, D.A. & Norman, R.W. (1980). Transfers of mechanical energy within the total body and mechanical efficiency during treadmill walking. *Ergonomics* **23**, 147–56.

Pinshow, B., Fedak, M.A. & Schmidt-Nielsen, K. (1977). Terrestrial locomotion in penguins: it costs more to waddle. *Science* **195**, 592–4.

Pollock, M.L. (1977). Submaximal and maximal working capacity of elite distance runners. The marathon: physiological, medical, epidemiological and psychological studies. *Ann. N.Y. Acad. Sci.* **301**, 310–22.

Prestwich, K. N. (1983). Anaerobic metabolism in spiders. *Physiol. Zool.* **56**, 112–21.

Rall, J.A. (1986). Energetic aspects of skeletal muscle contraction: implication of fiber types. *Exer. and Sports Sci. Rev.* **13** (R.L. Terjung ed.), Macmillan, New York, pp. 33–74.

Schmidt-Nielsen, K. (1972). Locomotion: energetic cost of swimming, flying and running. *Science* **177**, 222–8.

Seeherman, H.J., Taylor, C.R., Maloiy, G.M.O. & Armstrong, R.B. (1981). Design of the mammalian respiratory system: measuring maximum aerobic capacity. *Respir. Physiol.* **44**, 11–24.

Seeherman, H.J., Dmi'el, R. & Gleeson, T.T. (1983). Oxygen consumption and lactate production in varanid and iguanid lizards: a mammalian relationship. *Int. Series on Sport Sci.* **13**, 421–7.

Stainsby, W.N. & Fales, J.T. (1973). Oxygen consumption for isometric tetanic contractions of dog skeletal muscle *in situ*. *Am. J. Physiol.* **224**, 687–91.

Stainsby, W.N., Gladden, L.B., Barclay, J.K. & Wilson, B.A. (1980). Exercise efficiency; validity of base-line subtractions. *J. Appl. Physiol.* **48**, 518–22.

Taylor, C.R. (1977). The energetics of terrestrial locomotion and body size in vertebrates. In: *Scale Effects in Animal Locomotion* (ed. T.J. Pedly), Academic Press, London, pp. 111–25.

Taylor, C.R. (1980). Mechanical efficiency: a useful concept? In: *Aspects of Animal Movement* (H.Y. Elder & E.R. Trueman eds.), Cambridge University Press, Cambridge, pp. 235–44.

Taylor, C.R. (1985). Force development during sustained locomotion: a determinant of gait, speed and metabolic power. *J. exp. Biol.* **115**, 253–62.

Taylor, C.R., Heglund, N.C. & Maloiy, G.M.O. (1982). Energetics and mechanics of terrestrial locomotion. I. Metabolic energy consumption as a function of speed and body size in birds and mammals. *J. exp. Biol.* **97**, 1–21.

Taylor, C.R., Heglund, N.C., McMahon, T.A. & Lonney, T.R. (1980). Energetic cost of generating muscular force during running: a comparison of large and small animals. *J. exp. Biol.* **86**, 9–18.

Taylor, C.R., Schmidt-Nielsen, K. & Raab, J.L. (1970). Scaling of energetic cost to body size in mammals. *Am. J. Physiol.* **210**, 1104–7.

Taylor, C.R., Shkolnick, A., D'meil, Baharav, D. & Borut, A. (1974). Running in cheetahs, gazelles, and goats: energy cost and limb configuration. *Am. J. Physiol.* **227**, 848–50.

Tucker, V. (1970). Energetic cost of locomotion in animals. *Comp. Biochem. Physiol.* **34**, 841–6.

Williams, K.R. (1985).The relationship between mechanical and physiological energy estimates. *Med. Sci. Sports Exerc.* **17**, 317–25.

Williams, K.R. & Cavanagh, P.R. (1983). A model for the calculation of mechanical power during distance running. *J. Biomechanics* **16**, 115–28.

Williams, T.M. (1983). Locomotion of the North American mink, a semi-aquatic mammal. II. The effect of an elongate body on running energetics. *J. exp. Biol.* **48**, 153–61.

Winter, D.A. (1979). *Biomechanics of Human Movement*, John Wiley & Sons, New York, p.84.

8

Respiration in air breathing vertebrates: optimization and efficiency in design and function

W. K. MILSOM

8.1 INTRODUCTION

Efficiency, loosely defined, describes the ease with which a system performs its requisite tasks. By its more strict definition, it describes the ratio of the work produced by a system to the work expended on the system in performing that task. Several papers in this collection argue that the term 'efficiency' should be reserved only for the latter case. The former definition is more appropriately referred to as the effectiveness of the system. Thus, the efficiency of a respiratory system will refer only to the ratio of the mechanical work produced by the respiratory pump to the oxidative cost of producing that work. The effectiveness of this performance will also be a function of the many factors which influence the amount of gas exchanged between the environment and the blood for any given level of ventilation. This too can be regarded in terms of an input/output relationship and thus the effectiveness of a respiratory system can be defined as the ratio of the millilitres of O_2 exchanged during ventilation to the millilitres of O_2 required to power ventilation. The respiratory systems found in vertebrates reflect multiple experiments in design which attempt to optimize gas exchange across the respiratory surfaces given both the respiratory and non-respiratory constraints placed on the systems. Adaptive changes can be found at all levels which contribute both to the effectiveness and efficiency of each system. In the following article, select examples will be used to illustrate some of the trends which optimize the diffusive and convective processes involved in gas exchange (anatomical adaptations and their mechanical

consequences) as well as the manner in which they are linked (physiological and behavioural adaptations).

8.2   ANATOMICAL ADAPTATIONS

8.2.1   Lung morphology

To meet effectively the constraints placed on the respiratory system
for gas exchange, lungs must be designed to match diffusing capacity
to metabolic demands. Metabolic demands, on the other hand, vary
greatly between vertebrate groups, particularly between the ectotherms
and endotherms, as well as within any given group as a function of
both life style and body size. The structure and function of vertebrate
gas exchangers have been extensively reviewed in a number of recent
publications but several points relevant to our discussion of effectiveness
and efficiency will be presented here.

The basic structure of the respiratory system of air-breathing vertebrates differs markedly between different groups. According to the
type classification system proposed by Duncker (1978), all mammals
possess a bronchoalveolar lung while all birds possess a parabronchial
type lung. Reptiles, on the other hand, possess a continuum of lung
structures which have been categorized as unicameral, paucicameral
and multicameral. The unicameral lung consists of an undivided single chamber in which the respiratory surfaces are elaborations of the
lung wall. In paucicameral lungs, the central lumen of the lung is subdi-
vided by a small number of large septa while in the multicameral lung,
a cartilage-reinforced, intrapulmonary bronchus connects a number of
separate chambers (Duncker, 1978).

Perry (1983) has recently reviewed the trends which exist in lung morphology among the reptiles, birds and mammals. Lung volumes vary
extensively in size among the vertebrates, from less than 1.0 to over 100
ml/100 g (Fig. 8.1). Although there is much variation within each group,
in general, the lungs of reptiles tend to be larger than those of birds
which in turn are larger than those of mammals. Within each structural
type of lung, larger lungs usually exhibit a more heterogeneous distribution of parenchyma than smaller lungs. These trends, both within
and between groups, to some extent reflect constraints placed upon the
lungs which are not associated with increasing gas exchange surface *per
se.* Large, flexible, sac-like regions of the lungs of various vertebrates

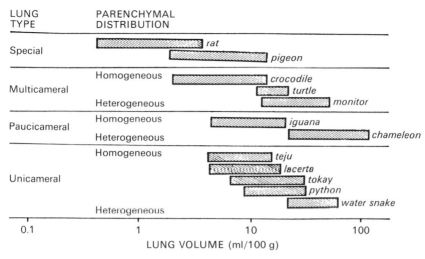

Fig. 8.1. Lung volumes in various reptiles, rat and pigeon (lung and air sacs), grouped according to lung structure. The lower limit of each bar indicates resting lung volume, the upper limit, vital capacity. (Modified from Perry, 1983).

play putative roles in gas storage, buoyancy control, locomotion, communication and behavioural displays (Perry, 1983). The constraints imposed by the need for gas exchange will be most clearly reflected in the anatomical diffusion factor (ADF) of the lung. This variable represents the anatomically measurable component of the diffusing capacity of lung tissue (Dt_{O2}) for oxygen and is equal to the ratio of the respiratory surface area to the harmonic mean thickness of the air-blood tissue barrier. The graphical representation of ADF shown in Fig. 8.2 (from Perry, 1983) clearly depicts the trend of increasing respiratory surface area (per gram body weight), decreasing air-blood tissue barrier thickness, and increasing anatomical diffusion factor as one progresses from reptiles through mammals to birds. It is of interest to note that in many instances, different strategies of increasing surface area or decreasing barrier thickness are employed to attain similar levels of ADF (compare chicken, rat and horse, for instance). Although the significance of these differences is not at all clear, there is a definite trend for both increasing surface area and decreasing barrier thickness with increasing metabolic rate. In Fig. 8.3, the morphometrically determined diffusing capacity of the air-blood tissue barrier is plotted against metabolic rate for many of the same species as in Fig. 8.2. This Figure, also taken from the work

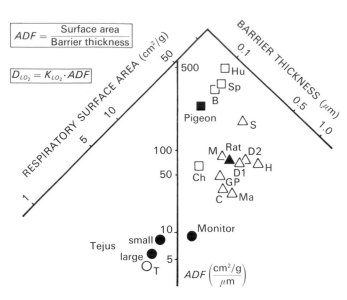

Fig. 8.2. Respiratory surface area plotted against the harmonic mean thickness of the air-blood tissue barrier. The quotient of these two parameters, the air diffusion factor (ADF), can be read off against the vertical axis. $\bigcirc$ = reptiles, $\triangle$ = mammals, $\square$ = birds. Darkened symbols represent animals in a similar body weight range (approximately 55 g). Abbreviations: reptiles: T = turtle; mammals: C = cow, D = dog, GP = guinea pig, H = horse, M = mouse, Ma = man, R = rat, S = shrew; birds: B = budgerigar, Ch = chicken, Hu = hummingbird, Sp = sparrow. (From Perry, 1983.)

of Perry (1983), clearly illustrates the correlation which exists between increasing metabolic demand (both resting and maximal) and the diffusing capacity of the lungs. Although the diffusing capacities plotted here are morphometric and not physiological and are all calculated using the Krogh's diffusion constant for oxygen of the rat lung (this constant has not been measured for most of the species depicted here), the trends which such calculations reveal are probably quite valid. The slope of the relationship shown here ($1/Pt_{O2}$) has the units torr. Although this measure is not identical with the physiologically determined alveolar-arterial partial pressure difference, it is probably, nonetheless, a good indicator of alveolar-arterial P_{O2} differences (see Perry, 1983, for discussion). This index shows a clear increase in the degree of equilibration between arterial blood and alveolar gas, under resting conditions again progressing from reptiles through mammals to birds, which is reduced, but still present, even under conditions of maximal oxygen uptake.

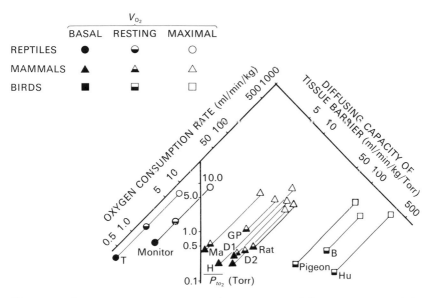

Fig. 8.3. Oxygen consumption rate plotted against diffusing capacity of the tissue barrier. The quotient of these two parameters (Pt_{O2}) represents the partial pressure difference between air and blood in the lung and is read off against the vertical axis. Values for reptiles, mammals and birds are given under basal and/or resting and maximal conditions. (From Perry, 1983, abbreviations as in Fig. 8.2.)

Just as differences in life style and metabolic rate among the various vertebrate groups lead to changes in lung structure which match diffusing capacity to metabolic demands, similar changes can be seen within any single group as a function of differences in weight specific metabolic rate which arise as a simple consequence of differences in body size. Kleiber (1961) and many others have demonstrated that for all vertebrates, oxygen consumption varies roughly with the three quarters power of body weight. As a consequence, one would expect that respiratory variables related to gas exchange should also scale as the three quarters power of body weight. Extensive literature reviews by several authors indicate that this is so (Zeuthen, 1953; Hemmingsen, 1960; Stahl, 1967; Hutchison *et al.*, 1968). Thus we see in Table 8.1 that for mammals, the total amount of gas ventilated each minute, the alveolar surface area available for gas exchange, the perfusion of the lungs, and the minute work or power required to move the gas all vary in direct proportion to metabolic rate. Interestingly, size (rather than function) related vari-

Table 8.1. *Respiratory variables in mammals expressed as allometric equations*

Variable	Units	Allometric Equation Body mass	Allometric Equation Metabolic rate
Oxygen consumption	ml/min	$11.6\ M^{0.76}$	—
Minute volume	ml/min	$379\ M^{0.80}$	$6.4\ MR^{1.06}$
Pulmonary surface area*	m^2	$131\ M^{0.76}$	$2.22\ MR^{1.00}$
Power of breathing	g.cm/min	$962\ M^{0.78}$	$16.4\ MR^{1.03}$
Cardiac output	ml/min	$187\ M^{0.81}$	$3.2\ MR^{1.07}$
Total lung weight	g	$11.3\ M^{0.99}$	
Vital capacity	ml	$56.7\ M^{1.03}$	
Tidal volume	ml	$7.69\ M^{1.04}$	
Dead space volume	ml	$2.76\ M^{0.96}$	

M = Body mass in kg; MR = metabolic rate in ml O_2/min.
* From Tenney and Remmers, 1963; all other values from Stahl, 1967.

ables of the respiratory system scale as the first power of body weight. As a consequence, such variables as lung weight, total lung volume, deadspace volume and tidal volume remain the same relative to body weight in mammals of all sizes. The physiological consequences of these trends have been reviewed elsewhere (Piiper and Scheid, 1972; Piiper, 1982). Pertinent to our discussion, and as pointed out by Tenney and Remmers (1963), as mammals decrease in size, their lung volume will decrease in direct proportion to their body weight while the internal surface area of the lung decreases in direct proportion to metabolic rate (or the three quarters power of body weight). One consequence of these two trends is that there must be an increase in the internal partitioning of the lung and thus a decrease in alveolar dimensions. Similar trends have now been described for several of these variables in birds, reptiles and amphibia (Hutchison *et al.*, 1968; Tenney and Tenney, 1970; Perry, 1983).

What all of these trends demonstrate, both within and between groups, is a decrease in lung volume per unit body weight but an increase in lung partitioning and complexity associated with increased weight specific metabolic demands. These adaptations, which all increase the effectiveness of the gas exchange surface, have a large influence on the mechanics of the repiratory system and thus, as we will see in the next section, on the efficiency of respiration.

8.2.2   *Respiratory pumps*

The gas exchange surfaces of most vertebrates are situated at protected sites within the body requiring respiratory pumps which must be designed to provide sufficient ventilation to match metabolic demands. The design of these pumps, however, is even more heavily influenced by other physiological, environmental and behavioural constraints than the gas exchange surfaces themselves. A consideration of the consequences of the long thin body of a snake or the rigid shell of a turtle on the design of the respiratory pump, the hydrostatic pressures encountered by diving animals on the structure of the lungs and chest wall, and the modifications associated with vocalization on the development of the lungs and respiratory passages in many species indicate the magnitude of these constraints. The literature is rich with the details of comparative studies of respiratory structure and function in various vertebrates outlining the diversity of solutions which exist to the different sets of constraints imposed on different species (see Milsom, 1989a, for a partial review). The consequences of these solutions on pulmonary mechanics have also been reviewed recently (Milsom, 1989b) and pertinent trends will be presented in the next section.

In recent years, much research has focused on the limiting factors which determine the maximum rate of oxygen consumption ($\dot{V}_{O2max}$) possible in any given species during intense exercise. It is currently believed that either tissue perfusion, capillary-tissue diffusion or, more likely, an interaction of these two processes determines $\dot{V}_{O2max}$ (Wagner, 1988). Except during severe exercise or under unusual conditions (altitude, for instance) or pathological circumstances, neither ventilation nor lung-pulmonary capillary diffusion are ever limiting. Thus the effectiveness of the respiratory systems of vertebrates appears to be more than adequate for even the maximum demands placed on the system.

8.3   MECHANICAL CONSEQUENCES

Adaptive changes in the structure of the gas exchange organs and respiratory pumps will not only influence the effectiveness of ventilation but will also alter pulmonary mechanics and, thus, ultimately will effect the efficiency of the respiratory pump. These changes in pulmonary mechanics can be viewed in terms of their effects on both the static and dynamic properties of the respiratory system.

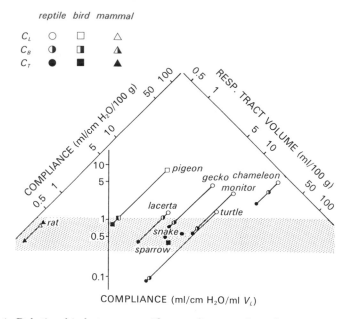

Fig. 8.4. Relationship between specific compliance and respiratory tract volume in seven reptilian, two avian and one mammalian species. The quotient of these two parameters is the volume-standardized compliance (C/V_L) which is read off against the vertical axis. C_L, C_B and C_T are the compliances of the lungs (and air sacs in birds), body wall and total system respectively. The stippled area indicates the range of C_T/V_L values found in most species. (Modified from Perry, 1983.)

8.3.1   Static mechanics

Effects on static mechanics are primarily reflected in the compliance (an indication of stiffness) of the respiratory system. Figure 8.4 illustrates the relative contributions of the compliance of the lungs alone and body wall alone to the total compliance of the respiratory system in different vertebrate species (from Perry and Duncker, 1980). This is a double logarithmic plot and both compliance and respiratory tract volume are normalized to body weight. When expressed this way, the lungs, body wall and total respiratory system of birds and reptiles appear to be from one to three orders of magnitude more compliant than those of mammals. The quotient of these variables (the slope of the relationship between weight specific compliance and lung volume) is the specific compliance (C/V_L) of the respiratory system and can be read off against the vertical axis in the figure. Since it is the function of the respiratory system which is under consideration, it seems more logical to normalize mechanical and functional variables to the size of the respiratory system rather than to the size of the animal. When expressed in this fashion it can be seen that the body wall compliance of most vertebrates is remarkably similar. The turtle, with its rigid shell, depends on movement of the walls of the flank cavities and pectoral girdle for volume change (McCutcheon, 1943, Gans and Hughes, 1967) and, not surprisingly, has a specific body wall compliance an order of magnitude stiffer than that of most other vertebrates. There is a large degree of variation in the specific compliance of the lungs of different species which to some extent reflects variations in lung architecture and lung volume (Perry and Duncker, 1980). The large, heterogeneous lungs of reptiles and lung-air sac systems of birds are much more compliant than the relatively small homogeneous lungs of mammals. Although the logarithmic scales used in Fig. 8.4 mask a 25-fold range of values from that of the turtle (and most mammals, not shown here) to that of the chameleon, specific compliance is also quite consistent between vertebrate classes despite the tremendous diversity in body structure and lung architecture found between these groups. It should be noted, however, that despite this relatively narrow range of values there is one fundamental difference. In mammals, C_T is determined primarily by the specific compliance of the lungs while in the birds and reptiles, C_T is determined primarily by the specific compliance of the body wall.

Given these similarities in C_T in all groups, we also find that all systems, regardless of species (from amphibians to mammals) or body size (up to seven orders of magnitude) can be inflated by similar changes in transpulmonary pressure (Fig. 8.5) (Agostini *et al.*, 1959; Milsom, 1989a). This suggests there may be real limits to the extent of transpulmonary pressure swings which can be utilized to power ventilation but the exact nature of the limiting factors remains unknown. That aside, the same change in transpulmonary pressure of two to five cm H_2O is required to expand the lungs by one tidal volume in each species, by over 50 L in a whale but under 0.1 ml in a shrew (Stahl, 1967; Leith, 1976).

8.3.2 Dynamic mechanics

During ventilation, dynamic forces are applied to the system which are met by equal and opposing forces developed by the system (Newton's

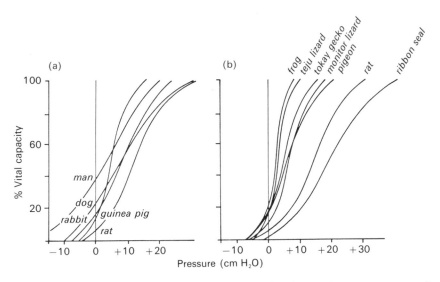

Fig. 8.5. Mean static pressure-volume curves for the respiratory systems of several species. (Mammalian data (a) from Agostoni *et al.*, 1959, and Leith, 1976; bird and reptile data (b) from Perry and Duncker, 1980; frog data from Hughes and Vergara, 1978.)

third law of motion). For the respiratory system, the applied pressure is opposed by pressures related to the volume, flow and volume acceleration of the respiratory system. The mechanical properties related to these pressures are the dynamic compliance, flow resistance and the inertance.

The compliances of the lung and respiratory system described above are affected by the rate of inflation or frequency of ventilation; the higher the ventilation frequency, the stiffer the behaviour of the system. Thus the dynamic compliance during ventilation is less than the corresponding compliance obtained under static conditions. In most mammals, this reduction in compliance is quite small (10% reduction over a 10-fold increase in ventilation frequency) (Sullivan and Mortola, 1987). In reptiles, however, it can be quite large (70% reduction over a six to 10-fold increase in ventilation frequency) (Vitalis and Milsom, 1986; Milsom and Vitalis, 1984), suggesting that stress relaxation in reptilian respiratory systems greatly exceeds that found in mammalian systems although the reasons for this are not clear.

The magnitude of the flow resistive forces which must be overcome during ventilation are a function of the flow resistance and the rate of movement of gas and tissues. These in turn are largely a function of the dimensions of the respiratory system. The flow resistance of the airways will decrease as lung volume increases because the airway dimensions increase as discussed earlier (Crosfill and Widdicombe, 1961; Tenney and Remmers, 1963; Rodarte and Rehder, 1986). The flow resistance of lung tissue and the chest wall should decrease with increasing lung volume because the pressure associated with tissue viscous resistance is proportional to the linear velocity of tissue and for a given volume change, the linear movement of tissue decreases as the volume increases (Grimsby *et al.*, 1968). One would also expect that flow resistance would be less in heterogeneous lungs relative to homogeneous lungs of equal size as a result of the large expansible chambers. Thus, with the trend of increasing homogeneity and decreasing alveolar dimensions with increasing weight specific metabolic rate described earlier, one also sees a good correlation between total flow resistance and weight specific metabolic rate.

8.3.3 *Mechanical cost of breathing*

In generating pressure, and hence flow, in the respiratory system, the respiratory muscles perform work to overcome the forces associated with the dynamic mechanical properties we have just discussed. In general, at low respiratory flow rates most of this work is required to overcome the elastic forces resisting lung inflation which will be a function of the tidal volume and the compliance of the system (Agostini *et al.*, 1970; Ruossos and Campbell, 1986). At higher respiratory flow rates, flow-resistive forces play a larger role in determining the total mechanical cost of ventilation. The sites and nature of the forces responsible for the generation of this work, however, differ dramatically from species to species. In Fig. 8.6, the curves labelled 'total' represent the relationship between the total cost of ventilating the entire respiratory system and the respiratory frequency at a constant level of pulmonary ventilation in a lizard, a turtle and a mammal. The work required to inflate the lungs as a percentage of the total work of breathing increases in the sequence gecko < turtle < mammal. This hierarchy also reflects the increasing complexity of lung architecture (Perry, 1983). Given the simple structure of their lungs, the total work of breathing in the Tokay gecko is almost entirely required to overcome elastic forces arising from the body wall at all frequencies and tidal volumes (Milsom and Vitalis, 1984). In turtles, the mechanical work required to inflate the respiratory system is required to overcome both elastic and non-elastic forces over

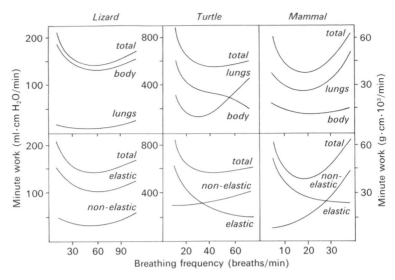

Fig. 8.6. Upper panels: The minute work required to ventilate the lungs, body compartment and total respiratory system in a lizard, turtle and mammal, at a constant level of pulmonary ventilation, as a function of ventilation frequency. Lower panels: The amount of minute work required to overcome elastic and non-elastic forces in the total respiratory system, at a constant level of pulmonary ventilation, as a function of ventilation frequency in the same three species. Values for minute work for the lizard and turtle are in ml.cm H_2O/min (left-hand axis) while those for the mammal are in g.cm/min (right-hand axis). (Data from Milsom and Vitalis, 1984 (lizard); Vitalis and Milsom, 1986 (turtle); and Otis *et al.*, 1950 (mammal).)

the entire range of frequencies studied. At low frequencies, elastic forces predominate with these forces residing primarily in the body wall and cavity. The work required to overcome non-elastic forces, however, predominates at high frequencies with the lungs contributing an increasing portion of this work (Vitalis and Milsom, 1986). In mammals, the work required to inflate the respiratory system is primarily required to overcome elastic forces at low respiratory frequencies and non-elastic forces at high frequencies but in both instances, these forces reside primarily in the lungs (Otis *et al.*, 1950; Agostoni *et al.*, 1959).

There are very few measures of the total mechanical work of breathing in non-mammalin species. Calculations of the elastic work of resting breathing (Perry and Duncker, 1980) as well as some direct measures of total mechanical work of breathing (Milsom, 1984; Vitalis and Milsom, 1986; Table 8.2) suggest that, due to higher compliances, the relative cost

Table 8.2. *Cost and efficiency of breathing in different vertebrates*

Species	Oxidative cost (kcal/min/kg)	(% $\dot{V}_{O2}$)	(ml/L)	Mechanical cost (kcal/kg/min)	Efficiency (%)
Human	1.64×10^{-1}	1–2	0.5–1	3.74×10^{-4}	3–10
Dog	1.44×10^{-4}	0.6	0.2	2.13×10^{-5}	15
Turtle	5–20×10^{-4}	10–30	5–15	1.25×10^{-6}	0.1–0.25
Frog	4.06×10^{-4}	5	0.5–0.7	3.25×10^{-6}	8
Trout	2.49×10^{-4}	1–7	0.1–0.3	2.49×10^{-6}	0.6–1.2

(From Milsom 1989a)

of ventilation in reptiles and birds is one to three orders of magnitude lower than in mammals. Table 8.2 also shows that the total mechanical cost of ventilation in amphibia, which use a buccal force pump, and fish, which ventilate a much denser medium with a lower oxygen content, can be one to two orders of magnitude lower than that seen in mammals. Clearly, anatomical differences in lung structure and pump design effect mechanical work.

8.3.4   Oxidative cost of breathing and breathing efficiency

As with the total mechanical cost of breathing, there is not much data available for the oxidative cost of breathing in different vertebrate groups. Table 8.1 lists the oxidative costs for at least one member of each major group other than the birds. These values are presented as absolute values normalized to body weight, as millilitres of O_2 consumed per L of gas moved, and as a percentage of the resting metabolic rate. While the oxidative cost per L of gas moved is roughly an order of magnitude greater in the turtle than in mammals, the cost of ventilation in the frog is surprisingly low. Despite this, when differences in metabolic rate and minute ventilation are taken into account, the relative cost of ventilation as a fraction of the total oxygen consumption of the animal is an order of magnitude greater for all three species of lower vertebrate compared to the mammals.

Knowing the oxidative cost of ventilation as well as the mechanical cost, the efficiency of the respiratory pump can be calculated. Table 8.2 lists the calculated efficiencies for the same five species. Given the difficulties of measuring both the total mechanical work of breathing as well as the oxidative cost of breathing, these calculations are probably only rough indications of efficiency. They do indicate, however, that with the exception of the turtle, the efficiency of breathing is in the order of one to 15%. Overall, these efficiencies are quite low and the consequences for species with low resting metabolic rates may be significant. In mammals, the work of breathing is a small fraction of the total energy turnover of the resting animal. As a consequence, it is usually considered of minor physiological importance. The cost at higher levels of ventilation does become somewhat larger, but even so, not very significant (Otis, 1954). The data for the lower vertebrates presented here indicate that even at rest, the oxygen cost of breathing is a major fraction of their total metabolic rate. The oxidative cost during periods of strenuous activity could become prohibitive. This indicates the importance of physiological and behavioural strategies which optimize breathing patterns and minimize the work of breathing. These strategies are considered next.

8.4   PHYSIOLOGICAL ADAPTATIONS

8.4.1   Breathing pattern and work of breathing

It has been well established on theoretical grounds that in all vertebrates, at any given level of total ventilation, mechanical work decreases as respiratory frequency increases with a concomitant decrease in tidal volume (Roussos and Campbell, 1986). This stems from the fact that the work required to overcome elastic forces is inversely proportional to frequency while that required to overcome non-elastic forces is independent of frequency (Otis, 1954). The same, however, is not true of pulmonary ventilation. For any given level of pulmonary ventilation there is an optimum frequency at which the cost of ventilation is minimal (Otis *et al.*, 1950). As pointed out by Otis (1954), to maintain a constant level of pulmonary ventilation with increasing respiratory frequency and decreasing tidal volume, actually requires an increase in total ventilation. This is required to offset the proportionate increase in dead space ventilation. As a consequence, increasing frequency tends to increase elastic work because of the increase in total ventilation while at the same time it tends to decrease elastic work because of the decrease in tidal volume. It is because of these opposite effects of changing breathing pattern on elastic work that an optimum frequency occurs although non-elastic work also determines the exact nature of the relationship between minute work

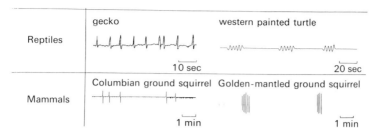

Fig. 8.7. Traces of pneumotachograph air flow during spontaneous breathing in two species of reptiles and in two species of mammals during hibernation, illustrating the basic patterns of arrhythmic breathing. (From Milsom, 1988.)

and breathing frequency (Otis, 1954; Roussos and Campbell, 1986). If we return to the relationships between the total work required to ventilate the respiratory system and breathing frequency, at a constant level of pulmonary ventilation, depicted in Fig. 8.6, one can see that there is an optimum frequency at which minute work is a minimum in each species. Despite the vast differences in the pulmonary mechanics of each group described earlier, the interaction between pulmonary mechanics and the relative costs of dead space ventilation invariably produce an optimal frequency.

The correspondence between these calculated and predicted optimal breathing frequencies and observed breathing frequencies, in all species which have been studied, is striking (Agostoni *et al.*, 1959; Crosfill and Widdicombe, 1961; Milsom and Vitalis, 1984; Vitalis and Milsom, 1986). It strongly suggests that pulmonary ventilation is regulated according to the principle of minimum effort as postulated by Rohrer (1925). It has been suggested that this correlation may be somewhat coincidental and that the frequency of spontaneous ventilation corresponds more closely with that associated with the minimum force required by the respiratory muscles (Mead, 1960). The difference in these predicted frequencies is not large and given the shape of the relationships between minute work or minimum force and respiratory frequency, there is very little difference in energy cost between breathing patterns utilizing either one (Otis, 1954). At present, the physiological mechanisms which underlie this phenomenon remain unclear although the data attests to it's universality.

The calculations of optimal tidal volume and respiratory frequency discussed above all assume a continuous breathing pattern with passive expiration (Otis, 1954; Roussos and Campbell, 1986). For several mammals and birds (diving species in particular), as well as the amphibians and reptiles, breathing is periodic and expiration is active (Milsom, 1987). The consequences of active expiration for predictions of optimal breathing pattern are small. Recent measurements of the mechanical cost of ventilation in spontaneously breathing lizards and turtles produce values which are not significantly different from those calculated from mechanical studies on pump-ventilated animals with passive expiration (Milsom, 1984; Morgan and Milsom, unpublished). Presumably the increased cost of active expiration was small and offset by stored energy resulting from the active expiration which would be available to help power inspiration. The consequences of periodic breathing for predictions of optimal breathing frequency, however, are another matter. There are two general patterns of periodic breathing commonly seen in these animals (Fig. 8.7). One consists of single breaths separated by relatively short periods of breath holding while the other consists of episodes of more or less continuous breathing separated by much longer periods of breath holding. On theoretical grounds, continuous breathing in the episodic breathers should correspond closely to predicted patterns based on the mechanical considerations discussed above. This is certainly the case for the turtle, *Pseudemys scripta* (Vitalis and Milsom, 1986). In the right hand panel of Fig. 8.8, the power required to maintain a constant level of pulmonary ventilation is plotted, along with that required to maintain a constant level of total ventilation, as a function of ventilatory frequency. Note that, due to dead space ventilation, minute work is always greater for any given level of pulmonary ventilation compared to a similar level of total ventilation and that the difference increases as frequency increases because of the proportionate increase in dead space ventilation along with rising non-elastic forces. The shaded area in this panel indicates the range of instantaneous breathing frequencies measured in spontaneously breathing animals and corresponds quite closely with the pump frequencies which require the minimum rate of work to maintain a constant level of pulmonary ventilation. They correspond less well to those frequencies at which the minimum work is required to maintain a constant level of total ventilation (Vitalis and Milsom, 1986).

The optimal breathing frequency for animals which take only single breaths followed by a variable pause, however, should be determined from an analysis of the work per breath rather than the minute work of breathing. Marine turtles and tortoises exhibit such patterns (Shel-

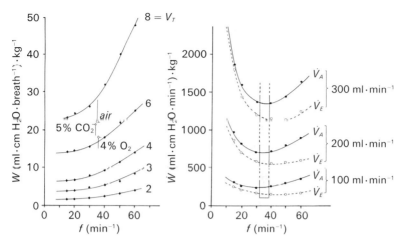

Fig. 8.8. Left-hand panel: The relationship between the work/breath and ventilation frequency for various levels of tidal volume (V_T) in the turtle (*Pseudemys scripta*). The open symbols represent the mean values $\pm$ SEM of V_T and instantaneous breathing frequency ($f' = 60/T_{Tot}$) measured from six animals spontaneously breathing air ($\bigcirc$), 5% CO_2 in air ($\triangle$), or 4% O_2 in N_2 ($\diamondsuit$). Right-hand panel: The minute work or power required to produce various constant levels of pulmonary ($\dot{V}_A$) and total ($\dot{V}_E$) ventilation as a function of breathing frequency. The stippled area represents the range of instantaneous breathing frequencies (f') measured in spontaneously breathing animals. (From Vitalis and Milsom, 1986.)

ton *et al.*, 1986; Milsom, 1987). and, if the pulmonary mechanics and respiratory variables of these species are similar to those of *Pseudemys scripta*, the work of breathing associated with such a pattern can be analyzed using the left-hand panel in Fig. 8.8. This panel displays the mechanical cost required to produce single breaths at various tidal volumes (V_T) as a function of the instantaneous ventilation frequency ($60/T_{Tot} = f'$, where $T_{Tot} =$ the total length of the active phases of a single breath). The values of V_T and f' for spontaneously breathing *P. scripta*, breathing various gas mixtures, are placed on the graph for comparison. These work-per-breath curves illustrate that for any given frequency, the smaller V_T, the lower the cost of each breath. A low V_T, however, will compromise pulmonary ventilation and gas exchange. Presumably, the need to keep V_T sufficiently large in order to maintain pulmonary ventilation, and the increased mechanical work associated with increases in V_T, interact to produce the resting level of V_T. At any given V_T, the work per breath decreases as breathing frequency decreases.

There are physiological and behavioural factors, however, which should restrict breath duration from becoming too long. A very long breath in a spontaneously breathing animal would require a slow, controlled expiration and inspiration which would require expiratory braking as well as dissipate the stored elastic energy which partially powers both expiration and inspiration. It would also tie up accessory respiratory muscles which frequently subserve other roles in reptiles (Gans, 1970) as well as reduce the breath hold period which would serve to restrict dive lengths in aquatic species. Given the shape of the relationship between work per breath and ventilatory frequency shown in Fig. 8.8, the decrease in mechanical work associated with a fall in ventilatory frequency below the f' measured in spontaneously breathing animals is very small compared with the large increase in the work per breath associated with an increase in breathing frequency above that value. The value of f' measured in spontaneously breathing animals may thus represent a compromise between mechanical and biological constraints.

With increasing metabolic demands, animals which exhibit periodic breathing patterns can increase the tidal volume of each breath, increase the speed of the breath or simply reduce the total period of breath holding by taking more breaths per minute. The data presented here would suggest that it is less costly to restrict changes in V_T and f' and take more breaths of a similar length and depth (in an episode or individually) and thus shorten the non-ventilatory period, than it is to increase V_T or shorten T_{Tot}. This is in fact what is seen in most reptiles (see Shelton *et al.*, 1986; Milsom, 1987, for reviews). The hypothesis would predict that a reduction in T_{Tot} would only occur when ventilation was sufficiently elevated that all periods of breath holding had disappeared.

8.4.2   *Breathing pattern and gas exchange*

It has been shown that for all vertebrates, lung blood flow is generally adjusted to match levels of pulmonary perfusion to levels of ventilation (Daly, 1986). This is particularly true of animals which exhibit periodic breathing patterns, giving rise to a 'ventilation tachycardia' (Johansen and Burggren, 1985). Given differences in gas exchange kinetics and sites of body stores, the effectiveness of this for O_2 and CO_2 exchange may be quite different as can be seen in Figs. 8.9 and 8.10. Figure 8.9 depicts the changes in the gas composition measured at the nose of a hibernating ground squirrel (*Spermophilus lateralis*) during a bout of

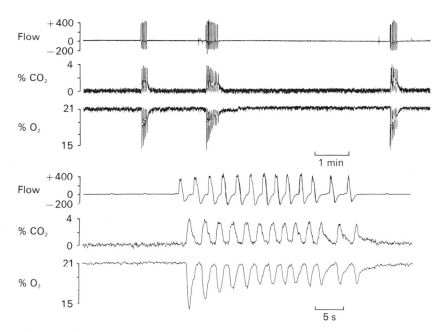

Fig. 8.9. Traces of pneumotachograph air flow and the fractional concentration of CO_2 and O_2 in expired air of a hibernating golden-mantled ground squirrel (7°C body temperature). The lower set of traces are on a faster time scale to illustrate the details of a single respiratory episode.

hibernation (7°C, body temperature). Of particular note is the fact that the end-expired O_2 concentration increases progressively throughout the breathing episode while the CO_2 concentration falls, but at a slower rate. Figure 8.10 shows the progressive increase in O_2 stores and fall in CO_2 stores throughout the episode depicted in the lower panel of Figure 8.9. Any slope through the origin on such a graph represents the rate of change in gas composition and thus the steeper the slope, the higher the rate of O_2 extraction or CO_2 excretion. These curves show that the rate of O_2 uptake falls off progressively after about the first four or five breaths in the episode. The rate of CO_2 excretion, however, increases over the first eight to 10 breaths, presumably as CO_2 stores are moved from the tissues to the lungs, and remains high until the last breath of the sequence. It thus appears that in this instance, cardio-respiratory coupling is able to maintain optimal (presumably) rates of CO_2 excretion over the entire respiratory episode.

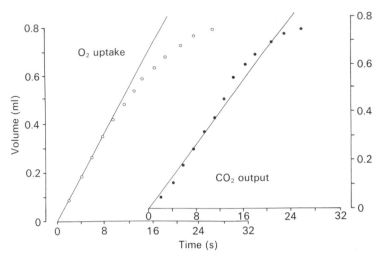

Fig. 8.10. The cumulative O_2 uptake (left axis) and CO_2 output (right axis) derived for each succesive breath (symbols) from the lower trace in Fig. 8.9.

8.4.3 Respiratory-locomotor coupling

Under situations of increased demand for gas exchange, the need for good ventilation-perfusion matching is also important. The metabolic scopes of many species exceed a 10-fold increase in metabolic rate requiring proportional increases in cardiac output and ventilation. The basic respiratory responses to metabolic demand, however, may differ significantly between exercise and other metabolic demands. Much of this stems from the recruitment of co-ordinated axial and appendicular mechanisms during exercise which in many instances, couple locomotor and respiratory rhythms. Such coupling has now been documented for many species, particularly those that use synchronous gaits (i.e. birds in flight, Hart and Roy, 1966; rabbits hopping, Viala *et al.*, 1987; wallabies bounding, Baudinette *et al.*, 1983; horses and dogs galloping, Bramble and Carrier, 1983; Carrier, 1987; Bramble, 1989) (Fig. 8.11). At present, little is known about the structural and biomechanical basis for respiratory-locomotor coupling or its physiological consequences. Is such coupling an adaptation or is it demanded by mechanical linkages between muscles of locomotion and the respiratory pump? Are these linkages adaptive?

The increased ventilatory demand put on the respiratory muscles dur-

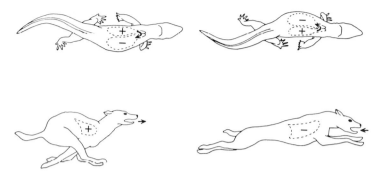

Fig. 8.11. Comparison of the effect that axial bending has on lung volume in a running lizard and a galloping dog. In the case of the lizard, the bending axis of the thorax is situated between the right and left lobes of the lungs. As the lizard bends laterally to one side, one lobe is expected to experience a reduction in volume while the other expands. Air may be pumped (indicated by arrows) back and forth between the lungs, but little if any will be moved in and out of the animal. In contrast, the bending axis of a galloping mammal is dorsal to the thoracic cavity. Sagittal bending changes the thoracic volume and actively pumps air in and out of the lungs during each locomotor cycle. (From Carrier, 1987.)

ing locomotion is quite large and the metabolic demands of these muscles will also increase reducing the net effectiveness of gas exchange in supplying the locomotor muscles. Given the relatively low efficiencies of respiratory pumps described earlier, it can be argued that many species take advantage of the locomotor movements, produced by more efficient muscles (20–25%, Margaria, 1976), to help power ventilation. It should be noted that the net work done by the animal remains the same (now the locomotor muscles are being used to power both ventilation and motion) but the net cost of locomotion (and the required level of gas exchange) should be reduced. At present, however, this is just conjecture as there is really insufficient data to support any hypothesis. Respiratory-locomotor coupling is not observed during exercise in all species and, in fact, there are instances where the mechanical linkage between respiratory and locomotor muscle constrains respiration during exercise. In many reptiles, the muscles of the respiratory pump also subserve other functions, including locomotion. In lizards where locomotion employs a sprawling posture and lateral vertebral bending (and hence a synchronous contraction of the muscles along one entire side of the thorax and abdomen (including both inspiratory and expiratory muscles) and relaxation of those on the other side of the body) (Fig. 8.11), ventilation is constrained. Minute ventilation decreases at higher running speeds and may be temporarily disrupted coincident with locomotor movements (Carrier, 1987).

8.5 CONCLUSIONS

In any discussion of optimization, one must be careful to avoid circular and/or teleological arguments. In an evolutionary context, one must assume that systems are as efficient as constraints allow. Examination of costs and efficiency only serve to give us a better understanding of the constraints placed on the system and allow us to view the design of these systems both within the context of any specific function as well as in the broader context of the net demands being placed on the system. The overall efficacy of respiratory systems in vertebrate animals is influenced by the anatomy of the gas exchange organs, the mechanics of the respiratory pumps and their pattern of ventilation. Due to both respiratory and non-respiratory constraints, vertebrates exhibit tremendous diversity in each of these factors. Despite this diversity, anatomical, mechanical, physiological and behavioural adaptations interact to improve gas exchange while reducing oxidative and mechanical costs and increasing efficiency. These adaptive trends are particularly important in lower vertebrates where the oxidative costs of resting ventilation may account for up to one third of the resting metabolic rate.

8.6 ACKNOWLEDGEMENTS

This work was supported by the NSERC of Canada. I am particularly grateful to Marianne Voldstedlund for collecting the data which appears in Figs. 8.9 and 8.10.

8.7 REFERENCES

Agostini, E., Campbell, E.J.M. & Freeman, S. (1970). Energetics. In: *The Respiratory Muscles* E.J.M. Campbell, E. Agostoni & J. Newsom Davis (eds.). Lloyd-Luke Ltd., London.

Agostini, E., Thimm, F.F. & Fenn, W.O. (1959). Comparative features of the mechanics of breathing. *J. Appl. Physiol.* **14**, 679–83.

Baudinette, R.V., Gannon, B.J., Runciman, W.B., Wells, S. & Love, J.B. (1987). Do cardiorespiratory frequencies show entrainment with hopping in the tammar wallabby? *J. exp. Biol.* **129**, 251–63.

Bramble, D.M. & Carrier, D.R. (1983). Running and breathing in mammals. *Science* **219**, 251–6.

Bramble, D.M. (1989). Axial-appendicular dynamics and the integration of breathing and gait in mammals. *Am. Zool.* **29**, 171–86.

Carrier, D.M. (1987). The evolution of locomotor stamina in tetrapods: circumventing a mechanical constraint. *Paleobiology* **13**, 326–41

Crosfill, M.L. & Widdicombe, J.G. (1961). Physical characteristics of the chest and lungs and the work of breathing in different mammalian species. *J. Physiol.* **158**, 1–14.

Daly, M. De B. (1986). Interactions between respiration and circulation. In: *Handbook of Physiology.* Section 3. The Respiratory System. Volume II. Control of Breathing, Part 2. A.P. Fishman, N.S. Cherniack, J.G. Widdicombe & S.R. Geiger (eds.). American Physiological Society, Bethesda.

Duncker, H.R. (1978). Funktionsmorphologie des Atemapparates und Coelom-Gliederung bei Reptilien, Vogeln unf Saugern. *Verh. Dtsch. Zool. Ges.* **1978**, 99–132.

Gans, C. (1970). Strategy and sequence in the evolution of the external gas exchangers of ectothermal vertebrates. *Forma et Functio* **3**, 61–104.

Gans, C. & Hughes, G.M. (1967). The mechanism of lung ventilation in the tortoise, *Testudo graeca* Linne. *J. exp. Biol.* **47**, 1–20.

Grimsby, G., Takishima, T., Graham, W., Macklem, P. & Mead, J. (1968). Frequency dependence of flow resistance in patients with obstructive lung disease. *J. Clin. Invest.* **47**, 1455–65.

Hart, J.S. & Roy, O.Z. (1966). Respiratory and cardiac responses to flight in pigeons. *Physiol. Zool.* **30**, 201–306.

Hemmingsen, A.M. (1960). Energy metabolism as related to body size and respiratory surfaces and its evolution. *Reports of the Steno. Mem. Hosp. and the Nord. Insulinlaboratorium* **9**, 1–110.

Hutchison, V.H., Whitford, W.G. & Kohl, M. (1968). Relation of body size and surface area to gas exchange in anurans. *Physiol. Zool.* **41**, 65–85.

Johansen, K. & Burggren, W.W. (1985). *Cardiovascular Shunts.* Alfred Benzon Symposium 21. Munksgaard, Copenhagen.

Kleiber, M. (1961). *The Fire of Life.* John Wiley & Sons, New York.

Leith, D.E. (1976). Comparative mammalian respiratory mechanics. *Physiologist* **19**, 485–510.

Margaria, R. (1976). *Biomechanics and Energetics of Muscular Exercise.* Clarendon, Oxford.

McCutcheon, F.H. (1943). The respiratory mechanism in turtles. *Physiol. Zool.* **16**, 255–69.

Mead, J. (1960). Control of respiratory frequency. *J. Appl. Physiol.* **15**, 325–7.

Milsom, W.K. (1984). The interrelationship between pulmonary mechanics and the spontaneous breathing pattern in the Tokay lizard, *Gekko gecko. J. exp. Biol.* **113**, 203–14.

Milsom, W.K. (1988). Control of arrhythmic breathing in aerial breathers. *Can.*

J. Zool. **66**, 99–108.

Milsom, W.K. (1989a). Mechanics of breathing; comparative aspects. In: *Lung Biology in Health and Disease; Volume 39: Comparative Pulmonary Physiology: Current Concepts.* C. Lenfant & S.C. Wood (eds.). Marcel Dekker Inc., New York.

Milsom, W.K. (1989b). Mechanisms of ventilation in lower vertebrates: adaptations to respiratory and non-respiratory constraints. *Can. J. Zool.* **67**, 2943-55.

Milsom, W.K. & Vitalis, T.Z. (1984). Pulmonary mechanics and the work of breathing in the lizard, *Gekko gecko. J. exp. Biol.* **113**, 187–202.

Otis, A.B., Fenn, W.O. & Rahn, H. (1950). Mechanics of breathing in man. *J. Appl. Physiol.* **2**, 592–607.

Otis, A.B. (1954). The work of breathing. *Physiol. Rev.* **34**, 449–58.

Perry, S.F. (1983). Reptilian lungs. *Adv. Anat. Embryol. Cell Biol.* **79**, 1–81.

Perry, S.F. & Duncker, H.R. (1980). Interrelationship of static mechanical factors and anatomical structure in lung evolution. *J. Comp. Physiol.* **138**, 321–34.

Piiper, J. (1982). Respiratory gas exchange at lungs, gills and tissues: mechanisms and adjustments. *J. exp. Biol.* **100**, 5–22.

Piiper, J. & Scheid, P. (1972). Maximum gas transfer efficacy of models of fish gills, avian lungs and mammalian lungs. *Respir. Physiol.* **14**, 115–24.

Rodarte, J.R. & Rehder, K. (1986). Dynamics of respiration. In: *Handbook of Physiology. Section 3. The Respiratory System. Vol. III. Mechanics of Breathing, Part 1.* A.P. Fishman, P.T. Macklem, J. Mead & S.R. Geiger (eds.). American Physiological Society, Bethesda.

Rohrer, F. (1925). Physiologie der Atembewegung. In: *Handbuch der Normalen und Path. Physiologie.* Vol. 2. A.T.J. Bethe, G. von Bergmann, G. Embden & A. Ellinger (eds.). Springer-Verlag, Berlin.

Ruossos, C. & Campbell, E.J.M. (1986). Respiratory muscle energetics. In: *Handbook of Physiology. Section 3. The Respiratory System. Vol. III. Mechanics of Breathing, Part 1.* A.P. Fishman, P.T. Macklem, J. Mead & S.R. Geiger (eds.). American Physiological Society, Bethesda.

Shelton, G., Jones, D.R. & Milsom, W.K. (1986). Control of breathing in ectothermic vertebrates. In: *Handbook of Physiology. Section 3. The Respiratory System. Volume II. Control of Breathing, Part 2.* A.P. Fishman, N.S. Cherniack, J.G. Widdicombe & S.R. Geiger (eds.). American Physiological Society, Bethesda.

Stahl, W.R. (1967). Scaling of respiratory variables in mammals. *J. Appl. Physiol.* **22**, 453–60.

Sullivan, K.J. & Mortola, J.P. (1987). Age related changes in the rate of stress relaxation within the rat respiratory system. *Respir. Physiol.* **67**, 295–309.

Tenney, S.M. & Remmers, J.E. (1963). Comparative quantitative morphology of the mammalian lung: diffusion area. *Nature* **197**, 54–6.

Tenney, S.M. & Tenney, J.B. (1970). Quantitative morphology of cold-blooded lungs: Amphibia and Reptilia. *Respir. Physiol.* **9**, 197–215.

Viala, D., Persegol, L. & Palisses, R. (1987). Relationship between phrenic and extensor activities during fictive locomotion. *Neuroscience Letters* **74**, 49–52.

Vitalis, T.Z. & Milsom, W.K. (1986). Pulmonary mechanics and the work of breathing in the semi-aquatic turtle, *Pseudemys scripta. J. exp. Biol.* **125**, 137–55.

Wagner, P.D. (1988). An integrated view of the determinants of maximum oxygen uptake. In: *Oxygen Transfer From Atmosphere To Tissues.* N.C. Gonzalez & M.R. Fedde (eds.). Plenum, New York.

Zeuthen, E. (1953). Oxygen uptake as related to body size in organisms. *Quart. Rev. Biol.* **28**, 1–12.

9

Cardiac energetics and the design of vertebrate arterial systems

D. R. JONES

Efficiency is the quotient of external work divided by total energy transformed and, in keeping with most biological systems, the efficiency of the vertebrate heart is low, ranging from 10–20%. Hence, only a small proportion of cardiac oxygen consumption appears as external work and functional or design strategies which improve external work performance will have relatively little impact on cardiac energetics. The majority of cardiac oxygen uptake is utilized for developing tension, generating the blood pressure, so anything which decreases cardiac tension (e.g. a fall in ventricular volume or blood pressure) or the time over which it is generated (e.g. a decrease in heart rate) will bring about a marked improvement in cardiac efficiency (Sarnoff et al., 1958). Given the physiological necessity for maintenance of high blood pressure, particularly in birds and mammals, it is possible to identify features of the evolutionary design of arterial systems which reduce the tension-time index and thereby improve cardiac efficiency.

The 'windkessel' or lumped-parameter model represents the simplest approach to describing the vertebrate arterial tree. The major arteries are considered to function as a single elastic reservoir which feeds into a single non-compliant resistance element (Fig. 9.1). When the heart contracts blood flows into the expanding elastic chamber ($\dot{Q}_c$) as well as through the peripheral resistance ($\dot{Q}_r$) (Fig. 9.1). During diastole, the elastic recoil of the stretched arterial walls, bearing down on the contained blood, delivers blood ($\dot{Q}_c$) to the peripheral resistance. Consequently, a highly pulsatile flow into the arterial system is transformed into a fairly steady outflow. The transformation in the flow pulse is also reflected in the pressure pulse for pulsations around the mean pressure,

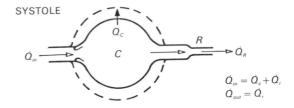

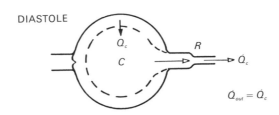

Fig. 9.1. A 'windkessel' model of the circulation. The major arteries are represented by a single elastic chamber with compliance C, while the peripheral vasculature is represented by a single resistance R. Total flow in systole, $\dot{Q}_{in}$, is the sum of the volume change of the elastic chamber, Q_c, and the simultaneous flow across R, $\dot{Q}_r$. Flow in diastole ($\dot{Q}_c$) is caused by the recoil of the stretched walls of the elastic chamber (figure and caption modified from Shadwick *et al.*, 1987).

at any given level of resistance, are reduced as the compliance increases. The arterial 'windkessel' creates steady flow in the capillaries ensuring body cells receive a continuous supply of substrates; however its role in reducing peak systolic pressures and thereby the tension-time index of the heart is frequently overlooked. Increases in input frequency to the 'windkessel' will also reduce pulsatility about the mean pressure but here energetic savings are compromised by the increase in the time that the cardiac muscle is generating tension.

Arteries function as elastic reservoirs in all vertebrate arterial systems but not all vertebrate arterial systems function as a simple 'windkessel'. A 'windkessel' model implies that pressure changes generated by cardiac contraction occur simultaneously throughout the arterial system (McDonald, 1974). However, each heart beat sends out a pulse wave which travels through the arterial system and arrives later at sites more distal to the heart. When the pulse transit time occupies a considerable proportion of each cardiac cycle (at least 10%) then significant phase

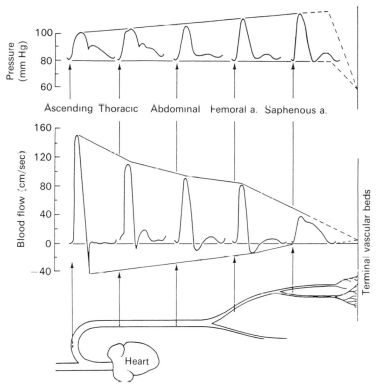

Fig. 9.2. Changes in pressure and flow velocity pulses from the ascending aorta to the saphenous artery. Oscillatory pressure increases, and oscillatory flow decreases, with increasing distance from the aorta. The broken lines indicate the decline in oscillatory pressure and flow that occur in distributing vessels leading to the microcirculation (redrawn incorporating modifications made by Folkow and Neil (1971) to an original in McDonald, 1960).

changes occur between the pressure and flow pulses at different arterial sites leading to marked deviations between pulses expected in a 'windkessel' and those actually recorded. For instance, the systemic pressure pulse exhibits marked changes in waveform as it travels through the arterial system. The pulse is amplified and secondary waves appear in the diastolic portion (Fig. 9.2).

By and large, the discrepancies between a 'windkessel' and real life are due to wave transmission effects with wave reflection being paramount. For example, pressure and flow pulses should look quite similar when recorded at the same point in a reflectionless system. Recordings of pressure flows at various sites along the systemic arterial system in birds and mammals show that this is clearly not the case (Fig. 9.2, McDonald, 1974; Langille and Jones, 1975; Noordergraaf *et al.*, 1979).

Analysis of wave propagation through the arterial system is complex and involves harmonic analyses of pressure and flow wave forms. Fortunately, a simple conceptual analysis of the interaction of incident and reflected waves in the arterial tree is sufficient for present purposes. All forms of wave motion can be reflected by any change in the system they are travelling through. At such changes within the arterial system incident pressure and flow waves will be reflected back towards the heart. The changes can be discrete discontinuities, due to arterial branching (McDonald, 1974), or continuous variations in elasticity due to the arteries stiffening as the periphery is approached (Langille & Jones, 1976). However, the major reflecting site seems to be the terminal vascular bed. Hence, the pressure and flow pulse waves are reflected back towards the heart and interfere, destructively or constructively, with the incident wave generated by cardiac contraction.

A major question concerns the nature of the termination that the peripheral vascular beds present to outgoing pressure and flow waves. Peripheral beds are 'closed' terminations if they present a relatively large impedance to pulsatile flow, they are 'open' if they present a relatively low impedance. In higher vertebrates, reflections produce large oscillations in peripheral pressures that drive small oscillatory flows through the terminal vascular beds (Fig. 9.2), indicating a high terminal impedance, i.e. of the 'closed' type. Hence, the pressure should be 100% reflected at the closed end without a phase shift while, to satisfy the condition that high pressure oscillations are required to drive low oscillatory flows through a high terminal impedance, the reflected flow wave is inverted, i.e. the reflected wave is 180° out of phase with the incident wave. However, how much of the incident wave reaches the termination (because the pulse is attenuated, especially in the smaller vessels) and how much of the reflected wave gets back to the heart (it too being reduced by damping) is a matter of speculation.

It has been argued that reflection effects have a beneficial effect on cardiac efficiency over and above that conferred by the 'windkessel' function of the arterial tree. For simplicity, consider a pressure wave displaying a simple harmonic motion illustrated in Fig. 9.3. At the closed end of the system (terminal vascular beds) both incident and reflected pressure waves will be in phase and the waves will sum giving an antinode, an enlarged pressure oscillation. The reflected wave is shown as 40% of the

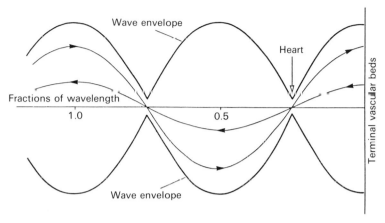

Fig. 9.3. A diagram to show the interaction of an incident and a reflected pressure wave at a closed end (terminal vascular beds). For clarity it is assumed that only 40 % of the wave is reflected (i.e. reflection coefficient = 0.4). The abscissa is marked in fractions of a wavelength. At the point of reflection both waves are in phase and sum together. With reference to a point one quarter wavelength away, the incident wave is 90° earlier and the reflected wave 90° later so that they are 180° out of phase and cancel. This point is a node and the only oscillation is the difference between the maximum amplitudes of the incident and reflected waves. For maximum benefits in terms of promoting cardiac efficiency the heart should be located at this point. The total excursion throughout the cycle is represented by the wave envelope (heavy outer lines), (figure redrawn and caption modified from McDonald, 1974).

incident wave in Fig. 9.3 so that the resultant, compound, wave will be 140% of the incident wave as is shown by the wave envelope (Fig. 9.3). Now consider a point a quarter wavelength (¼) back from the termination. The incident wave left here one quarter of a cycle before it reached the closed end and the reflected wave takes another quarter cycle to get back to this point, so the incident and reflected waves are now 180° out of phase and destructive interference produces a node. Hence, the resultant wave is some 40% smaller (as shown by the wave envelope in Fig. 9.3) than it would be in the absence of reflections. It has been estimated that, at the level of the aorta, some 20–40% of the arterial pressure pulse may be due to reflections (Frank, 1905, 1930; Li, 1986; Westerhoff *et al.*, 1972), so the tension generated by the ventricle is reduced substantially.

The flow wave is reflected such that the incident and reflected waves are 180° out of phase at the termination but are in phase one quarter cycle away from the terminal impedance. Hence, the waves will cancel one another at the termination and will add to one another at the heart. Therefore, highly oscillatory cardiac outflow is generated for less pulsatile pressure and the external work of the heart is increased, at any given level of cardiac oxygen consumption (Milnor, 1979). Furthermore, a reduced pulse pressure implies that lower systolic pressures generate the same mean pressure, so the tension-time integral is also reduced. Thus beneficial effects of reflections will be maximized when the heart is 0.25 λ upstream of the major reflecting site (Fig. 9.3). In birds and mammals the pulse wave velocity varies from 4 m s^{-1} in the aorta to 10 m s^{-1} in the peripheral vasculature. Assuming, an average velocity of 5–6 m·s^{-1} (O'Rourke *et al.*, 1968) then the major reflecting site in man (a basketball player perhaps?) will be up to 1.5 m from the heart at resting heart rates. However, at maximum heart rates (220 minus age in beats·min^{-1}), the reflection site will appear to be only 0.5 m from the heart, i.e. the abdominal aorta. In other words, maximal energy savings from reflections are more likely to occur during exercise and not at rest as was suggested by Noordergraff *et al.* (1979).

Heart rate should decrease with increase in body size in proportion to the change in linear dimension (i.e. $M_b^{-0.33}$) if the heart is to remain 0.25 λ back from the major reflecting site in birds and mammals. This is certainly not the case for resting animals. The relationship for birds is $f_h = 155.8 \, M_b^{-0.23}$ (Hartman, 1955; Calder, 1968) and for mammals $f_h = 241 \, M_b^{-0.25}$ (Brody, 1945; Stahl, 1967), where f_h = resting heart frequency (min^{-1}) and M_b is body mass in kg. The exponents for both birds and mammals are considerably below the -0.33 expected. Furthermore, the discrepancy will not be improved during exercise because heart rate generally increases less in small than in large animals. Hence, on present evidence, it is not reasonable to draw the conclusion that transmission effects offer a major design constraint on the arterial system. However, it is obvious that wave transmission effects offer savings in cardiac energetics even in an imperfectly designed arterial system. In general, energetic advantages accrue with tachycardia but are lost during bradycardia. Many marine mammals, in fact, spend much of their lives with heart rate way below those of their terrestrial counterparts. Substantial bradycardia develops during submergence regardless of whether diving is voluntary or forced. At a heart rate of 30 min^{-1}, the pulse wavelength is 16 m or so, and pulse transmission time is insignificant compared with the cycle period. Reflections will no longer be effective in reducing either the tension-time index or external work. Diving animals, however, have improved on the basic 'windkessel', for in many, the aortic root is greatly expanded to form the aortic bulb and the consequences of such an expansion on cardiac energetics have been modeled by Campbell *et al.* (1981). Campbell *et al.* (1981) built a mathematical model of the arterial system of the dog which, in computer simulations, gave similar pressure and flow waveforms to those recorded experimentally in various parts of the canine arterial system. This model was modified to resemble the diving seal by increasing peripheral resistance and increasing the radius of the ascending aorta by 2.26 times while retaining the radius to wall thickness ratio constant. The total added distensibility (C_{tot}) varied from 1.27 to 3.43 times the distensibility of the original dog model. At a heart rate of $10 \cdot \text{min}^{-1}$ and with a dive-type increase in peripheral resistance, the unmodified dog model generated a peak systolic pressure of 156 mm Hg and a minimum diastolic pressure of 63.6 mm Hg. However, addition of extra compliance in the form of an aortic bulb resulted in peak systolic and diastolic pressures of 112.5 and 70.5 mm Hg respectively, representing considerable energy savings in terms of the tension-time index. Campbell *et al.* (1981) also experimented with the location of the added compliance (Fig. 9.4a). A large central compliance (I) was more effective in reducing peak systolic pressure than similar compliances put in the brachiocephalic and thoracic aorta (II), the last two segments of the thoracic aorta only (III), and the terminal vascular beds (IV) (Fig. 9.4a). However, all locations were equally effective in raising diastolic pressure (Fig. 9.4b). Consequently, the heart sees all of the compliance added by increasing the radius of the ascending aorta while much of the added compliance is hidden from the heart at other vascular locations.

The strategy seen in the seal recapitulates that adopted by many lower vertebrates which have a greatly expanded arterial segment just downstream of the heart. In teleost fish this segment is elastic (*bulbus arteriosus*) whereas in elasmobranchs and amphibians the structure is both elastic and contractile (*conus arteriosus*). The proximal arterial segment will reduce systolic pressure levels (as shown in the seal model) and thereby reduce cardiac oxygen consumption. Also, the large increase in proximal arterial distensibility will ensure maintained flow throughout the cardiac cycle, even just outside the bulbus.

The majority of fishes and amphibians are small and have low heart rates so wave transmission time through the arterial system will take less than 5% of the cardiac cycle (Langille and Jones, 1977). This is

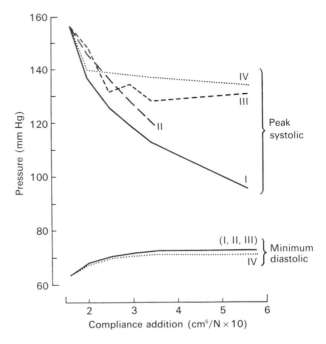

Fig. 9.4. Effects of compliance addition to the canine arterial system on peak systolic pressure development and minimum diastolic pressure in the aorta. The canine model was modified to reflect changes occurring in a diving seal, i.e. bradycardia and increased vascular resistance. Compliance addition was achieved in four different distribution patterns, as described in the text, and separate effects from each pattern are labeled I, II, III and IV, respectively (redrawn from Campbell *et al.*, 1981).

insufficient for the development of marked phase differences between pressure and flow waves and wave interference phenomena will not be apparent. However, some elasmobranch and teleost fishes can grow quite spectacularly and may have high heart rates ($> 100 \text{ min}^{-1}$) so, although wave transmission phenomena are unlikely in the short ventral aorta before the gills, it is possible that wave transmission effects could be seen in the dorsal circulation between the gills and systemic capillaries (Langille *et al.*, 1980). The role that would be played by dorsal aortic reflections is difficult to estimate although it seems unlikely that they would be seen by the heart.

From a haemodynamic viewpoint the larger reptiles appear to bridge the gap between the lower and higher vertebrates. The bulbar region of the aorta is obliterated by backwards growth of walls of the aortic arches but, even so, in many of these reptiles the base of the aorta is expanded relative to the size of vessels that arise from them, outside the pericardium. This might be viewed as a reptilian equivalent of the aortic bulb. However, in larger elongate reptiles this 'windkessel insurance' may be of little consequence because at heart rates of 60 min^{-1} and body lengths of three–six meters the stage appears to set for pronounced reflection effects. Unfortunately, we have virtually no information on arterial haemodynamics of large reptiles although studies in the area should be most interesting and particularly fruitful.

9.1 ACKNOWLEDGEMENTS

I am grateful to Dr B. Lowell Langille for his insightful comments while I was preparing the manuscript and an extremely critical, but tremendously helpful review of an early draft. Dr W.K. Milsom made a number of suggestions which improved the final draft of the manuscript. Also, I would like to thank Dr G.R.J. Gabbott for drawing the figures. My research cited in this review was supported by NSERC and the B.C. and Yukon Heart Foundation.

9.2 REFERENCES

Brody, S. (1945). *Bioenergetics and Growth, with Special Reference to the Efficiency Complex in Domestic Animals*. Reinhold, New York. 1023 pp.

Calder, W.A. (1968). Respiratory and heart rates of birds at rest. *Condor* **70**, 358–65.

Campbell, K.B., Rhode, E.A., Cox, R.H., Hunter, W.C. and Noordergraaf, A. (1981). Functional consequences of expanded aortic bulb: a model study. *Am. J. Physiol.* **240**, R200-R210.

Folkow, B. and Neil, E. (1971). *Circulation*. Oxford University Press, London. 593 pp.

Frank, O. (1905). Der Puls in den Arterien. *Z. Biol.* **46**, 441–553.

Frank, O. (1930). Schätzung des Schlagvolumens des menschlichen Herzens auf Grund der Wellen- und Windkesseltheorie. *Z. Biol.* **90**, 405–9.

Hartman, F.A. (1955). Heart weight in birds. *Condor* **57**, 221–38.

Langille, B.L. and Jones, D.R. (1975). Central cardiovascular dynamics of ducks. *Am. J. Physiol.* **228**, 1856–61.

Langille, B.L. and Jones, D.R. (1976). Examination of elastic non-uniformity in the arterial system using a hydraulic model. *J. Biomechanics* **9**, 755–61.

Langille, B.L. and Jones, D.R. (1977). Dynamics of blood flow through the heart and arterial systems of anuran amphibia. *J. Exp. Biol.* **68**, 1–17.

Langille, B.L., Stevens, E.D. and Anantaraman, A. (1980). Cardiovascular and respiratory flow dynamics. In *Fish Mechanics*, ed. P.W. Webb, pp. 92–139. Academic Press, London and New York.

Li, J.K.-J. (1986). Time domain resolution of forward and reflected waves in the aorta. *Trans. Biomed. Eng.* **33**, 783–5.

McDonald, O.A. (1960). *Blood Flow in Arteries.* Arnold, London.

McDonald, D.A. (1974). *Blood Flow in Arteries.* Williams and Wilkins, Baltimore. 496 pp.

Milnor, W.R. (1979). Aortic wavelength as a determinant of the relation between heart rate and body size in mammals. *Am. J. Physiol.* **237**, R3-R6.

Noordergraaf, A., Li, J.K.-J. and Campbell, K.B. (1979). Mammalian hemodynamics: A new similarity principle. *J. Theor. Biol.* **79**, 485–9.

O'Rourke, M.F. (1967). Pressure and flow waves in systemic arteries and the anatomical design of the arterial system. *J. Appl. Physiol.* **23**, 139–49.

Sarnoff, S.J., Braunwald E., Welch, G.H. Jr., Case, R.B. Stainsby, W. and Macruz, R. (1958). Hemodynamic determinants of oxygen consumption of the heart with special reference to the tension-time index. *Am. J. Physiol.* **192**, 148–56.

Shadwick, R.E., Gosline, J.M. and Milsom, W.K. (1987). Arterial haemodynamics in the cephalopod mollusc, *Octopus dofleini. J. Exp. Biol.* **130**, 87–106.

Stahl, W.R. (1967). Scaling of respiratory variables in mammals. *J. Appl. Physiol.* **22**, 53–60.

Westerhof, N., Sipkema, P., Van Den Bos, G.C. and Elzinga, G.(1972). Forward and backward waves in the arterial system. *Cardiovasc. Res.* **6**, 648–56.

10

An evolutionary perspective on the concept of efficiency: how does function evolve?

G. V. LAUDER

10.1 ABSTRACT

The concept of efficiency is an explicit and useful vehicle for comparing the functioning of organisms and their component parts, and as such is an invaluable comparative yardstick that enables us (1) to assess function quantitatively, (2) to compare dissimilar taxa and physiological systems, and (3) to analyze the evolution of physiological systems. Efficiency is defined as the ratio of work output to chemical energy input. Efficiency should not be confused with the concepts of performance or effectiveness which measure the ability of organisms to execute behaviors. In many cases performance is a more appropriate measure of function at the organ or whole organism level than efficiency where appropriate inputs and outputs may be hard to define.

From an evolutionary perspective, efficiencies should be measured relative to an outgroup clade, not merely in comparison to the most easily obtainable species. Three major questions in the evolution of function are posed and a general phylogenetic method for investigating these questions is considered. First, how well integrated are organisms: to what extent is the functioning of subsystems at various levels of organization within organisms matched? Secondly, is there a tradeoff between efficiency and the ability to regulate function in organisms? Thirdly, does evolution optimize efficiency, and what are the evolutionary consequences of changes in efficiency? There is a tendency to believe that changes in efficiency may be causally related to evolutionary 'success', but considerable caution is needed in assigning historical importance to a presumed change in efficiency: historical hypotheses of functional significance are difficult to test. The integration of quantitative analyses of organismal function with comparative and phylogenetic analyses will contribute to our understanding of how physiological systems evolve.

10.2   INTRODUCTION

The major theme of this paper is that an historical approach to the analysis of efficiency in organisms has the potential for clarifying many issues of importance to physiologists. For the most part, investigators who have used the concept of efficiency to evaluate animal structure and function have focused on proximal questions: how efficient are locomotor muscles, how efficient are particular biochemical reactions, how efficient are different species at moving from point A to point B etc. (e.g., Hildebrand, 1974; Milic-Emili and Petit, 1960; Taylor, 1980; Vogel, 1988)? As is well illustrated by many of the papers in this volume, the efficiency of both whole animals and parts of animals can be measured, and such measurements can tell us a great deal about the function of specific physiological systems. In addition, it seems to have been accepted by physiologists that measuring efficiency is a useful way to evaluate animal function.

Despite an increasing use of the concept of efficiency to study physiological systems, there are still many key questions that might fruitfully be addressed by integrating historical and phylogenetic analyses with physiological studies of efficiency. With remarkably few exceptions, there is little intellectual cross-fertilization between investigators who are interested in phylogenetic patterns to organismal design and those who study the function of specific physiological systems. And yet such interactions may prove to be an area in which considerable progress might be made on several key questions of interest to both historical and functional biologists. Thus, I feel that we need to foster an integration of quantitative analyses of organismal function with comparative and phylogenetic analyses to better understand how physiological systems evolve. The three goals of this paper are: (1) to discuss the definition of efficiency and related concepts that might be useful for evaluating organismal function, (2) to define specific issues in the study of animal function (focusing on the concept of efficiency) that might profit from an historical analysis, and (3) to discuss methodologies for analyzing historical patterns to physiological function.

10.2.1 The definition of efficiency

One tenet of this paper is that the concept of efficiency is a useful and explicit vehicle for comparing the function of organisms and their component parts. As noted by Blake (Chapter 2), investigators have used the concept of efficiency in different ways and have often measured it differently. Here I would like to define how the concept of efficiency will be used in this paper and emphasize how efficiency differs from the concepts of 'performance' and 'effectiveness'.

When the efficiency of any specific subsystem in an organism is measured the most common definition of efficiency is as the ratio of work output to work input or to chemical energy input (usually multiplied by 100 to give a value in percent), and this is the definition I will use here (Blum, 1970; Hildebrand, 1974; Vogel, 1988). With this definition it is easy to understand how efficiency could be measured at several different levels of biological organization. For example, in a study of muscle function during locomotion, the efficiency of sarcomere shortening could be studied by measuring the ratio of the work done by sarcomere shortening to the quantity of ATP used. Or, the efficiency of whole animal locomotion could be calculated as the ratio of the work done in moving a unit mass a unit distance to the energy used in converting stored fats or carbohydrates to ATP. In this example, the intermediate levels of the muscle fiber, whole muscle, and musculoskeletal system are bypassed in favor of measuring efficiency at two disparate levels.

This example illustrates a potential problem in extending the concept of efficiency beyond the analysis of a specific subsystem within an organism: as many levels of organization are spanned in a measure of efficiency, it becomes increasingly difficult to determine what the relevant inputs and outputs to the efficiency ratio are. Should the denominator of the efficiency ratio be ATP used, oxygen consumed, carbohydrate metabolized or raw foodstuffs needed? Depending on the problems of measuring these quantities and the assumptions that go into the measurement, different determinations of efficiency could be obtained.

While the concept of efficiency is attractive for the analysis of animal function, a key use of the concept of efficiency in comparative analyses is that it provides the ability to compare the function of organisms that may vary widely in size and shape and that may possess few homologous morphological features in common. For example, in aquatic locomotion the concept of Froude Efficiency (Ef) is defined as the ratio of useful power produced during locomotion to the useful power plus the power lost to the fluid (O'Dor and Webber, 1986; Trueman, 1980). Measuring Ef allows us to compare the locomotor hydrodynamics of fish (a rainbow trout has an Ef in the range of 0.61 to 0.81) to that of a squid (with an Ef of 0.2 to 0.3; Alexander, 1977; O'Dor and Webber, 1986). The difference in Froude Efficiency between these taxa appears to be largely due to the pulsatile nature of squid locomotion in contrast to swimming by trout in which force is being nearly continuously applied to the fluid. By using a common currency such as Froude Efficiency to measure animal function, we gain the ability to make comparisons that would otherwise be both difficult and qualitative. The ability to compare divergent taxa quantitatively is not a trivial advantage of a general concept like efficiency. It is vital for comparative and historical analyses of organismal function that quantitative comparative analyses be feasible.

The concept of efficiency should not be confused with the notion of performance, the ability to execute a behavior (Arnold, 1983; Emerson and Diehl, 1980; Garland *et al.*, 1988; Reilly and Lauder, 1988; Wainwright, 1987). In many cases performance is a more appropriate measure of function at the whole organism level than efficiency, where appropriate inputs and outputs may be very hard to define. Reasonable performance measures might include the percent of time that a prey escapes from a predator, the number of prey captured per feeding attempt, maximum burst speed, or the distance moved in a fixed time interval. Performance measures the ability of an animal to execute a particular behavioral test or the effectiveness of an animal at accomplishing a task. Performance *per se* does not measure the energetic expenditure or the efficiency of a behavior under the definition used in this paper.

Measures of performance may also be applied to specific organs or physiological systems. For example, the strength of skeletal elements might be compared by measuring the ratio of strength to skeletal mass. The specific strength of bones may vary within an organism and between taxa, and provides a measure of skeletal effectiveness at supporting loads. Measurements of specific strength reflect skeletal performance or effectiveness, not efficiency.

10.2.2   *Major questions in the evolution of function*

In the last twenty years major changes have occurred in the way comparative and phylogenetic analyses are conducted (Brooks, 1984; Brooks and Wiley, 1988; Eldredge and Cracraft, 1980; Ridley, 1983; Wiley,

1981). The vast majority of such research has been concerned with phylogenetic patterns of structure and the transformation of morphology. Relatively little attention has been given to how function evolves and to historical changes in organismal function. Furthermore, many hypotheses about physiological function may be tested with a comparative and historical approach, and yet only a very few investigators have succeeded in integrating physiological and historical analyses (e.g. Huey and Bennett, 1987).

Since the area of the evolution of function has received so little attention, there are many important questions that could be addressed, but as yet few answers. Here I would like to focus on three issues that are particularly relevant to the concept of efficiency, and use the analysis of efficiency to consider methodologies for investigating historical changes in organismal function.

(1) How well integrated are organisms: to what extent is the functioning of subsystems at various levels of organization within organisms matched? The issue of how well integrated organisms are is one of the oldest in biology. The nineteenth century morphologist Cuvier clearly expressed the view that organisms are highly integrated and that all parts are intercorrelated and must function as a coordinated whole. Cuvier states that ... all the organs of an animal form a single system, the parts of which hang together, and act and re-act upon one another; and no modifications can appear in one part without bringing about corresponding modifications in all the rest (Russell, 1982, p. 35). Indeed, Cuvier's view of the high degree of functional integration in complex morphological systems might even be seen as an indirect intellectual antecedent to the concept of 'symmorphosis' proposed by Weibel and Taylor (1981). Weibel (1984, p. 60), for example, in a discussion of the respiratory system, notes that 'in a functional system as complex as the one we are considering there is important interaction and cross-influence from one part to the other.' The concept of 'symmorphosis' suggests that structures at several levels of organization will be closely matched to their functional roles, and that there is a high degree of integration among levels. The issue of organismal integration is an important one in analyses of the evolution of function as the extent to which functions and structures are highly integrated might be expected to reflect the degree to which evolutionary changes may occur independently at different levels of design.

The question of how tightly linked changes at different levels of orga-

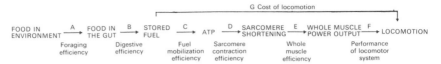

Fig. 10.1. Schematic diagram of a series of steps in the transformation of food in the environment to locomotor activity by an organism. The efficiencies of five processes (A to E) are shown. Step F represents measurement of locomotor performance and step G indicates measurement of the cost of locomotion (e.g., Daniel and Webb, 1987). A key question in the evolution of function is the extent of congruent change in phylogeny among the efficiencies at each link in this functional chain.

nization are is one that can be approached empirically in an historical context. If, for example, we wished to study the efficiency of locomotion, we might define a series of physiological processes or steps from the acquisition of food in the environment to movement across the ground (Fig. 10.1). The efficiency of each step in the chain can be measured to obtain a picture of the efficiency with which food is transformed to stored fuel, ATP is converted into length changes in sarcomeres, sarcomere contraction is converted into muscle length changes, etc. (Fig. 10.1: steps A to G).

In order to ask a question about the evolution of efficiency and the nature of the relationship among efficiencies at these different levels of organization, we need to measure efficiencies at several levels in several species. The analysis of one species alone does not tell us anything about the transformation of function in an historical sense. In addition, data from several species, even when analyzed in a traditional allometric fashion, also fail to inform us about patterns of historical change as allometric analyses assume that all points (species) are independent when in fact they are not (Felsenstein, 1985). This critical point is discussed in more detail below. First, let us consider the efficiences of several different functions in several species.

If we had data on the efficiencies of energy conversion at levels C, D, E, and G (Fig. 10.1) from several species (Fig. 10.2: species a to n), then we could map the distribution of each efficiency onto a phylogeny to examine how evolutionary changes at each level have occurred (Fig. 10.2). The central idea in the historical analysis of function is the examination of congruent change among variables on a phylogeny. The phylogeny provides the historical context while the extent of congruent change provides the measure of evolutionary association. If, for exam-

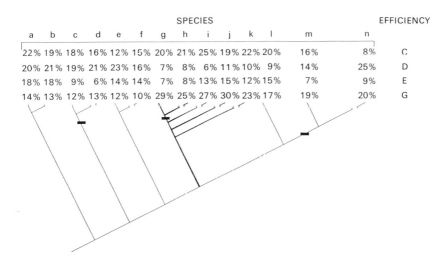

SPECIES   EFFICIENCY

a	b	c	d	e	f	g	h	i	j	k	l	m	n	
22%	19%	18%	16%	12%	15%	20%	21%	25%	19%	22%	20%	16%	8%	C
20%	21%	19%	21%	23%	16%	7%	8%	6%	11%	10%	9%	14%	25%	D
18%	18%	9%	6%	14%	14%	7%	8%	13%	15%	12%	15%	7%	9%	E
14%	13%	12%	13%	12%	10%	29%	25%	27%	30%	23%	17%	19%	20%	G

Fig. 10.2. Diagram of a cladogram representing the phylogeny of a group of species (a to n). Above each species the value of efficiency for each of three links in the chain shown in Fig. 10.1 is given as well as a measurement of the cost of locomotion. Note that the clade drawn in thick lines shares high values for the cost of locomotion and fuel mobilization efficiency, but low efficiencies of sarcomere contraction. The black bars on the stems leading to species c and d, g and h, and m and n represent the occurrence of these species in a novel habitat. No other species on this diagram are found in that habitat.

ple, changes in values of efficiency C (Figs. 10.1 and 10.2) always occur with changes in values of efficiency D or E, then we can say that, in the historical sense, these two levels of functional design are linked and have changed in concert. From the distribution of efficiencies in Fig. 10.2, one can see that one clade (shown in thick lines) is characterized by a higher cost of locomotion (G), a higher fuel mobilization efficiency (C), but a lower efficiency of sarcomere contraction (D) relative to other (outgroup) clades (such as c, d, e, and f; Fig. 10.2). This pattern indicates that there is no necessary positive physiological coupling between the cost of locomotion and sarcomere contraction efficiency. Since clade n in Fig. 10.2 exhibits both low fuel mobilization efficiency (C) and high sarcomere contraction efficiency (D) we can refute the hypothesis suggested by species g to l that these two efficiencies are necessarily inversely related.

Such a hierarchical analysis may provide insight into the causal basis of differences among species in behavior. If, for example, a clade in

Fig. 10.2 were found to have a relatively high cost of locomotion (and was therefore less efficient than other related clades) but was found also to have a high whole muscle efficiency (link E in Fig. 10.1), then one might wish to examine in detail the next link in the chain (i.e., link F in Fig. 10.1) for a possible explanation for this discrepancy. Perhaps those species with high costs of locomotion and high whole muscle efficiencies have an arrangement of the muscles and bones (such as shorter lever arms, less muscle mass, etc.) that produces the observed reduced performance.

In order to appreciate just how important a phylogeny is in interpreting the pattern of association between two physiological variables (such as those shown in Fig. 10.2), consider one alternative method of analyzing such data. If each species is treated as an independent data point and the data analyzed by correlation and linear regression analyses, one would estimate that the Pearson product-moment correlation (between the cost of locomotion (G) and efficiency D (Figs. 10.1, 10.2)) for these species to be -0.731 with a probability of 0.003. This would indicate a highly significant relationship between these two measures of physiological function. However, a scatterplot of these efficiencies shows that this relationship is determined largely by the species in one clade (Fig. 10.2: species g to k). Without this clade included (or with the mean value for the clade used), the correlation drops to -0.131 and is not significantly different from zero. The phylogeny illustrates how one could be mislead using a traditional allometric analysis: by assuming that all the species (including g to k) are independent, the regression degrees of freedom are greatly inflated. In fact, as the phylogeny of Fig. 10.2 illustrates, species g to k should be treated as one point since high values for G and low values for D evolved once in the ancestor of that clade.

Several methods have been proposed in the literature for quantitatively analyzing data such as those presented in Fig. 10.2, and these will not be considered in detail here (see, for example, Cheverud *et al.*, 1985; Felsenstein, 1985; Huey and Bennett, 1987; Losos, in press; Martins and Garland, in preparation; Swofford, 1984;). The general goal of these numerical analytical techniques is to allow reconstruction of states at the nodes on the phylogeny, and to provide a quantitative assessment of the extent of congruence among characters. As yet I know of no examples using the concept of efficiency where such a phylogenetic analysis has been attempted. The key message from a simple example such as that shown in Fig. 10.2 is that the related questions of organismal integration and functional evolution are empirical ones: via a combined phylogenetic and physiological analysis, considerable insight may be gained into the evolution of function.

Two subsidiary questions to the issue of organismal integration may also be addressed within a phylogenetic context. Firstly, do energetically demanding environments produce a close match of efficiencies at different levels of organization? For example, one might imagine that locomotion in a dense medium (such as water) might cause an increase in efficiency at each step in the chain of processes leading to work being done on the environment (as in Fig. 10.1). One way to ask such a question would be to add to Fig. 10.2 a measure of environmental demands on each species. One would then like to know the extent of historical congruence between changes in environment and changes in efficiency at each level. For example, three pairs of species in Fig. 10.2 exhibit low values of whole muscle efficiency (Fig. 10.2: species c, d; g, h; m, n). If it could be demonstrated that each pair of species also shares some aspect of the environment (indicated by the black bars in Fig. 10.2), then one would have demonstrated an historical correlation between the invasion of the novel environment and the presence of low whole muscle efficiency.

As an aside it is also important to note that the species exhibiting low whole muscle efficiency in Fig. 10.2 show only three independent evolutionary acquisitions of the novel habitat and muscle function. Although a total of six species are involved, only three independent historical origins have occurred. The use of six separate points in a graph showing a correlation between environment and muscle efficiency would be incorrect; there are three 'historical degrees of freedom'.

Secondly, does independence of efficiencies at different levels permit (in the historical sense) diversity of functions? In other words, is it true that the less tightly integrated two links in a chain such as that shown in Fig. 10.1 are, the greater the extent of divergence between the functions controlled by the those levels? Does a reduction in the level of integration permit independent specialization of function at each level, even though the overall efficiency of function across several levels may be less?

This question could be examined by comparing two clades that differ in the historical pattern of change in efficiencies. For example, consider one clade in which an historical analysis (such as that of Fig. 10.2) reveals that changes in the efficiencies of whole muscles in the limbs (link E in Fig. 10.1) are always associated with changes in sarcomere contraction efficiencies (link D in Fig. 10.1). This finding would suggest that there is a functionally important link that is not broken phylogenetically between these two levels. In contrast, a second clade of species (when subjected to a similar historical analysis) may exhibit considerable independence of change in links D and E. Is it true that this second clade exhibits, for example, a greater locomotor repertoire than species in the first clade?

The issue of the tradeoff between the tight integration of function among levels and flexibility of design has received little attention, but may be an important question in analyses of the evolution of function.

(2) Is there a tradeoff between efficiency and the ability to regulate function in organisms? The idea that there is a possible tradeoff of high efficiency at some particular level of design for other aspects of physiological function (such as power) has been discussed extensively in the literature (Gnaiger, 1987). However, to my knowledge there have been no attempts to test the historical hypothesis that regulatory ability and efficiency may make mutually conflicting demands on functional systems. From an evolutionary perspective this hypothesis could be tested by examining the level of congruence between some measure of regulatory ability (perhaps response to a perturbation) and efficiency at a particular level (such as whole muscle efficiency; level E in Fig. 10.1). The ability of an animal to modulate the activity of a particular biochemical pathway may be inversely proportional to the efficiency with which that pathway operates because of functional constraints imposed by the demands of high-efficiency output. While the analysis of biochemical efficiencies is widespread, I am not aware of any attempts to consider such a tradeoff from an historical perspective.

(3) Does evolution optimize efficiency, and what are the evolutionary consequences of changes in efficiency? There are many functional biologists who believe that evolution does optimize efficiency (or structures, functions, and behaviors) but the evidence from evolutionary biology does not support this position. Townsend and Calow (1981, p. 7) claim that 'We believe it can safely be assumed that natural selection will tend to produce organisms which are *maximally* effective at ... foraging, defending themselves, storing fat, growing, etc.' (italics mine), and Alexander (1982: p.1) states that 'Evolution by natural selection is a process of optimization.'

Such views of evolution stand in contrast to numerous analyses of both the process of evolution itself, the genetic basis of evolutionary change, and results from morphological and paleontological research (e.g., Cheverud, 1984; Gould, 1977, 1980a, b; Lewontin, 1974; Raup, 1972; Stearns, 1983). In particular, Lewontin (1983, p. 368) in his commentary entitled 'Elementary errors about evolution' states that 'it is simply factually incorrect to describe evolution as always being an adaptive or optimizing process.' Lewontin (1983) notes that (1) random events may result in the fixation of even deleterious genes in a population, (2) the environment is nearly always fluctuating (and is thus in a non-equilibrium state) which will cause features of organisms to be in a state of flux, and (3) genetic linkages among traits and pleiotrophic effects will cause selection on one character to cause a correlated response in other characters. Any of these three situations could produce structures, functions, or behaviors that are not close to some pre-defined optimal value. These observations, coupled with the fact that only very rarely have models of organismal function been successful in predicting performance without a proliferation of constraints and assumptions leads one to conclude that it must certainly be the exception rather than the rule when organismal design conforms to a previously defined optimal value. In addition, it is critical to realize that statements about natural selection and its role in shaping organismal design are nearly always assumptions invoked to justify the use of an optimality model, not a demonstrated fact (Cracraft, 1981).

Given that it is the rule, rather than the exception, that structures have multiple functions, and that both structures and functions are used by organisms to perform multiple biological roles (in the sense of Bock and von Wahlert, 1965; Gans, 1988) it is perhaps more useful to investigate patterns of functional interaction and how such patterns have changed than to take the procrustian approach of fitting organismal function into preconceived notions of how animals should be built.

I view the question of evolutionary increases in efficiency as an empirical one and urge that a priori assumptions about optimality or the nature of evolutionary change in physiology do not limit attempts to discover actual historical sequences of transformation. The analysis of efficiency within a phylogenetic context (e.g., Fig. 10.2) may show that, on average, the efficiency of some particular physiological system has increased over time. On the other hand, the efficiency of the same system may have decreased in other clades. Changes in efficiency that may be observed are certainly in no way a necessary product of the evolutionary process.

One notion that has achieved some currency in past discussions of the evolution of function is the idea that increases in efficiency are somehow responsible for the evolutionary 'success' of groups of animals. One example of this approach is the analysis of the evolution of locomotor function in ray-finned fishes. The transformation of the shape of the tail in ray-finned fishes is a textbook example of structural and functional change that is presented in nearly every comparative anatomy and paleontology text (e.g. Romer, 1966; Romer and Parsons, 1986; Carroll, 1988). Early ray-finned fishes possess a heterocercal tail in which the notochord extends into a dorsal lobe which is larger than the ventral lobe. When the tail is swept from side to side, the asymmetrical tail generates an epibatic (lift) force tending to cause the body to rotate about its center of mass. Lift forces to compensate for this rotation must be generated by other fins so that the fish can swim in a straight line. Teleost fishes are characterized by a novel (homocercal) tail structure in which the dorsal and ventral lobes are nearly equal in size (Gosline, 1971; Lauder, 1989; Marshall, 1971) and homocercal tend not to generate moments about the center of mass. The transformation from the primitive heterocercal condition to a derived homocercal tail has been taken to reflect an increase in locomotor efficiency that has had an important part in the origin of the 25,000 species of teleost fishes (Gosline, 1971; Lund, 1967). As summarized by Affleck (1950, p. 365), 'Because the fin of a homocercal tail swings about a vertical axis it is more efficient as part of the propulsive unit than the fins of a heterocercal tail'. Leaving aside the question of whether the efficiencies of the two types of tail have been actually measured, it may be proposed that an increase in efficiency of tail function at one phylogenetic level may lead to increased speciation and diversification in clades that possess the more efficient structure or function. The logical structure of such an argument has been considered in detail elsewhere (Lauder, 1981, 1982; Lauder and Liem, 1990) but it is critical to note that at each phylogenetic level there will be many characters that are uniquely associated with any given clade. In fish phylogeny, there are other characters that arise concordantly with the transformation of the tail. How are we to argue that the change in the tail alone (just one out of many changes) is causally related to speciation in ray-finned fishes. It is an easy error to assume that the origin of a feature or an increase in efficiency at one phylogenetic level is causally related to the subsequent history of a group, but our ability to test the historical effects of one such unique event is extremely limited (Lauder, 1981; Lauder and Liem, 1990), and such hypotheses often reflect our desire to see causal historical relationships where only a correlation has been demonstrated.

10.3 CONCLUSIONS

Historical analysis and hypothesis testing has only recently begun to achieve widespread attention, and the influx of historical analytical techniques and concepts into physiological analysis is still in its infancy. However, there is much to be gained by attempting a marriage of the disciplines of physiology and historical biology. The benefits of such cross-fertilization in other areas are clearly evident (for example, the liaison between neurobiology and ethology to produce the area of neuroethology has generated many new insights into organismal function). In physiology, we have only begun to understand the most general patterns to the evolution of function. Many questions have yet to be defined. In some systems, so few species have been studied that an historical or comparative analysis is just not possible. However, the concept of efficiency, used by so many workers on so many different physiological systems, offers us a potentially valuable avenue to begin to ask key questions about the evolution of function. The notion of efficiency, when precisely defined and applied, is an invaluable comparative yardstick that enables us to assess function quantitatively and to compare dissimilar taxa and physiological systems. Such comparisons form the foundation of any attempt to study the evolution of function.

10.4 ACKNOWLEDGEMENTS

I am particularly grateful to my colleagues who discussed aspects of the concept of efficiency with me: Grover Stephens, Al Bennett, Bob Josephson, Tim Bradley, Erich Gnaiger, Jean Malmud, Swifty Stephenson, Peter Wainwright, and Steve Reilly. This work was supported by NSF grants BSR 8520305 and DCB 8710210.

10.5 REFERENCES

Affleck, R. J. (1950). Some points in the function, development, and evolution of the tail in fishes. *Proc. Zool. Soc. Lond.* **120**, 349–68.

Alexander, R. McN. (1977). Swimming. In R. McN. Aexander and G. Goldspink (Eds), *Mechanics and Energetics of Animal Locomotion*, pp. 222–248 New York: John Wiley.

Alexander, R. McN. (1982). *Optima for Animals*. London: Edward Arnold.

Arnold, S. J. (1983). Morphology, performance, and fitness. *Amer. Zool.* **23**, 347–61.

Blum, H. F. (1970). *Time's Arrow and Evolution.* Princeton: Princeton Univ. Press.

Bock, W. and von Wahlert, G.(1965). Adaptation and the form-function complex. *Evolution* **19**, 269–299.

Brooks, D. R. (1984). What's going on in evolution? A brief guide to some new ideas in evolutionary theory. *Can. J. Zool.* **61**, 2637–45.

Brooks, D. R. and Wiley, E. O. (1988). *Evolution as Entropy*, 2nd Edn. Chicago: Univ. of Chicago Press.

Carroll, R. L. (1988). *Vertebrate Paleontology and Evolution.* New York: W. H. Freeman.

Cheverud, J. (1984). Quantitative genetics and developmental constraints on evolution by selection. *J. Theor. Biol.* **110**, 155–71.

Cheverud, J., Dow, M. M. and Leutenegger, W. (1985). The quantitative assessment of phylogenetic constraints in comparative analyses: sexual dimorphism in body weight among primates. *Evol.* **39**, 1335–51.

Cracraft, J. (1981). The use of functional and adaptive criteria in phylogenetic systematics. *Amer. Zool.* **21**, 21–36.

Daniel, T. L. and Webb, P. W. (1987). Physical determinants of locomotion. In *Comparative Physiology: Life in water and on land*, P. Dejours, L. Bolis, C. R. Taylor, and E. R. Weibel (Eds) pp. 343–369, Padova: Liviana Press.

Eldredge, N. and Cracraft, J.(1980). *Phylogenetic Patterns and the Evolutionary Process.* New York: Columbia Univ. Press.

Emerson, S. and Diehl, D. (1980). Toe pad morphology and mechanisms of sticking in frogs. *Biol. J. Linn. Soc.* **13**, 199–216.

Felsenstein, J. (1985). Phylogenies and the comparative method. *Amer. Nat.* **125**, 1–15.

Gans, C. (1988). Adaptation and the form-function relation. *Amer. Zool.* **28**, 681–97.

Garland, T., Geiser F. and Baudinette, R.V. (1988). Comparative locomotor performance of marsupial and placental mammals. *J. Zool., Lond.* **215**, 505–22.

Gnaiger, E. (1987). Optimum efficiencies of energy transformation in anoxic metabolism: the strategies of power and economy. In *Evolutionary Physiological Ecology*, P. Calow (Ed.), pp. 7–36. Cambridge: Cambridge Univ. Press.

Gosline, W. A. (1971). Functional morphology and classification of teleostean fishes. Honolulu: University Press of Hawaii.

Gould, S. J. (1977). *Ontogeny and Phylogeny.* Cambridge: Harvard Univ. Press.

Gould, S. J. (1980a). *The Panda's Thumb.* New York: W. W. Norton.

Gould, S. J. (1980b). The evolutionary biology of constraint. *Daedalus* **109**, 39–52.

Hildebrand, M. (1974). *Analysis of Vertebrate Structure.* New York: John Wiley.

Huey, R. B. and Bennett, A. F. (1987). Phylogenetic studies of coadaptation: preferred temperatures versus optimal performance temperatures of lizards. *Evolution* **41**, 1098–115.

Lauder, G. V. (1981). Form and function: structural analysis in evolutionary morphology. *Paleobiology* **7**, 430–42.

Lauder, G. V. (1982). Historical biology and the problem of design. *J. Theor. Biol.* **97**, 57–67.

Lauder, G. V. (1989). Caudal fin locomotion in ray-finned fishes: historical and functional analyses. *Amer. Zool.* **29**, 85–102.

Lauder, G. V. and Liem, K. F. (1990). The role of historical factors in the evolution of complex organismal functions. Chapter 5, pp. 63–78, In: *Complex Organismal Functions: Integration and Evolution in Vertebrates*, D. B. Wake and G. Roth, Eds. Dahlem Konferenzen, Chichester: John Wiley and Sons.

Lewontin, R. C. (1974). *The Genetic Basis of Evolutionary Change*. New York: Columbia Univ. Press.

Lewontin, R. C. (1983). Elementary errors about evolution. *Behav. Brain Sci.* **3**, 367–8.

Losos, J. in press. Concordant evolution of locomotor behavior, display rate and morphology in Anolis lizards. *Behavior*.

Lund, R. (1967). An analysis of the propulsive mechanisms of fishes, with references to some fossil actinopterygians. *Ann. Carneg. Mus.* **39**, 195–218.

Marshall, N. B. (1971). *Explorations in the Life of Fishes*. Cambridge: Harvard Univ. Press.

Milic-Emili, G. and Petit, J. M. (1960). Mechanical efficiency of breathing. *J. Appl. Physiol.* **15**, 359–62.

O'Dor, R. K. and Webber, D. M. (1986). The constraints on cephalopods: why squid aren't fish. *Can J. Zool.* **64**, 1591–605.

Raup, D. M. (1972). Approaches to morphologic analysis. In: T. Schopf (Ed.), *Models in Paleobiology*, pp. 28–44. San Francisco: W. H. Freeman.

Reilly, S. M. and Lauder, G. V. (1988). Ontogeny of aquatic feeding performance in the eastern newt, *Notophthalmus viridescens* (Salamandridae). *Copeia* **1988**, 87–91.

Ridley, M. (1983). *The Explanation of Organic Diversity*. Oxford: Clarendon Press.

Romer, A. S. (1966). *Vertebrate Paleontology*. Chicago: Univ. of Chicago Press.

Romer, A. S. and Parsons, T. S. (1986). *The Vertebrate Body*. Philadelphia: Saunders.

Russell, E. S. (1982). Form and function. A contribution to the history of animal morphology (Reprint). Chicago: Univ. of Chicago Press.

Stearns, S. (1983). The influence of size and phylogeny on patterns of covariation among life-history traits in the mammals. *Oikos* **41**, 173–87.

Swofford, D. L. (1984). *Phylogenetic Analysis Using Parsimony (PAUP), version 2.3.* Illinois Natural History Survey.

Taylor, C. R. (1980). *Mechanical efficiency of terrestrial locomotion: a useful concept animal movement.* H. Y. Elder and E. R. Trueman (Eds.), pp. 235-244. Cambridge: Cambridge Univ. Press.

Townsend, C. R. and P. Calow. (1981). *Physiological Ecology*. Sunderland: Sinauer Press.

Trueman, E. R. (1980). Swimming by jet propulsion. In: *Aspects of Animal Movement*. H. Y. Elder and E. R. Trueman (Eds.), pp. 93–105. Cambridge: Cambridge Univ. Press.

Vogel, S. (1988). *Life's Devices*. Princeton: Princeton Univ. Press.

Wainwright, P. C. (1987). Biomechanical limits to ecological performance: mollusc-crushing by the Caribbean hogfish, *Lachnolaimus maximus* (Labridae). *J. Zool., Lond.* **213**, 283–297.

Weibel, E. R. (1984). *The Pathway for Oxygen*. Cambridge: Harvard Univ. Press.

Weibel, E. R. and Taylor, C. R. (1981). Design of the mammalian respiratory system. *Respir. Physiol.* **44**, 1–164.

Wiley, E. O. (1981). *Phylogenetics*. New York: John Wiley.

INDEX